Lecture Notes in Computer Science 12444

More information about this series at http://www.springer.com/series/7412

Shadi Albarqouni · Spyridon Bakas ·
Konstantinos Kamnitsas ·
M. Jorge Cardoso · Bennett Landman ·
Wenqi Li · Fausto Milletari ·
Nicola Rieke · Holger Roth ·
Daguang Xu · Ziyue Xu (Eds.)

Domain Adaptation and Representation Transfer, and Distributed and Collaborative Learning

Second MICCAI Workshop, DART 2020
and First MICCAI Workshop, DCL 2020
Held in Conjunction with MICCAI 2020
Lima, Peru, October 4–8, 2020
Proceedings

 Springer

Editors
Shadi Albarqouni (iD)
Technical University Munich
Munich, Germany

Spyridon Bakas (iD)
University of Pennsylvania
Philadelphia, PA, USA

Konstantinos Kamnitsas (iD)
Imperial College London
London, UK

M. Jorge Cardoso (iD)
King's College London
London, UK

Bennett Landman (iD)
Vanderbilt University
Nashville, TN, USA

Wenqi Li
NVIDIA Ltd.
Cambridge, UK

Fausto Milletari (iD)
NVIDIA GmbH and Johnson & Johnson
Munich, Germany

Nicola Rieke (iD)
NVIDIA GmbH
Munich, Germany

Holger Roth
NVIDIA Corporation
Bethesda, MD, USA

Daguang Xu
NVIDIA Corporation
Bethesda, MD, USA

Ziyue Xu
NVIDIA Corporation
Santa Clara, CA, USA

ISSN 0302-9743   ISSN 1611-3349 (electronic)
Lecture Notes in Computer Science
ISBN 978-3-030-60547-6   ISBN 978-3-030-60548-3 (eBook)
https://doi.org/10.1007/978-3-030-60548-3

LNCS Sublibrary: SL6 – Image Processing, Computer Vision, Pattern Recognition, and Graphics

This Springer imprint is published by the registered company Springer Nature Switzerland AG
The registered company address is: Gewerbestrasse 11, 6330 Cham, Switzerland

Preface DART 2020

Computer vision and medical imaging have been revolutionized by the introduction of advanced machine learning and deep learning methodologies. Recent approaches have shown unprecedented performance gains in tasks such as segmentation, classification, detection, and registration. Although these results (obtained mainly on public datasets) represent important milestones for the MICCAI community, most methods lack generalization capabilities when presented with previously unseen situations (corner cases) or different input data domains. This limits clinical applicability of these innovative approaches and therefore diminishes their impact. Transfer learning, representation learning, and domain adaptation techniques have been used to tackle problems such as: model training using small datasets while obtaining generalizable representations; performing domain adaptation via few-shot learning; obtaining interpretable representations that are understood by humans; and leveraging knowledge learned from a particular domain to solve problems in another.

The second MICCAI workshop on Domain Adaptation and Representation Transfer (DART 2020) aimed at creating a discussion forum to compare, evaluate, and discuss methodological advancements and ideas that can improve the applicability of machine learning (ML)/deep learning (DL) approaches to clinical settings by making them robust and consistent across different domains.

During the second edition of DART, 18 papers were submitted for consideration and, after peer review, 12 full papers were accepted for presentation. Each paper was rigorously reviewed by three reviewers in a double-blind review process. The papers were automatically assigned to reviewers, taking into account and avoiding potential conflicts of interest and recent work collaborations between peers. Reviewers have been selected among the most prominent experts in the field from all over the world. Once the reviews were obtained the area chairs formulated final decisions over acceptance or rejection of each manuscript. These decisions were always taken according to the reviews and were unappealable.

Additionally, the workshop organization granted the Best Paper Award to the best submission presented at DART 2020. The Best Paper Award was assigned as a result of a secret voting procedure, where each member of the committee indicated two papers worthy of consideration for the award. The paper collecting the majority of votes was then chosen by the committee.

We believe that the paper selection process implemented during DART 2020, as well as the quality of the submissions, have resulted in scientifically validated and interesting contributions to the MICCAI community, and in particular to researchers working on domain adaptation and representation transfer.

We would therefore like to thank the authors for their contributions, the reviewers for their dedication and professionality in delivering expert opinions about the submissions, and the NVIDIA Corporation, which sponsored DART, for the support, resources, and help in organizing the workshop. The NVIDIA Corporation also sponsored the prize for the best paper at DART 2019, which consisted of a NVIDIA TITAN RTX GPU card.

August 2020

Shadi Albarqouni
M. Jorge Cardoso
Konstantinos Kamnitsas
Fausto Milletari
Nicola Rieke
Daguang Xu
Ziyue Xu

Preface DCL 2020

Machine learning approaches have demonstrated the capability of revolutionizing almost every application and every industry through the use of large amounts of data to capture and recognize patterns. A central topic in recent scientific debates has been how data is obtained and how it can be used without compromising user privacy. Industrial exploitation of machine learning and deep learning (DL) approaches, has on the one hand highlighted the need to capture user data from the field of application in order to yield a continuous improvement of the model, and on the other hand it has exposed a few shortcomings of current methods when it comes to privacy.

Innovation in the way data is captured, used, and managed, as well as how privacy and security of this data can be ensured, is a priority for the whole research community.

Most current methods rely on centralized data stores, which contain sensitive information, and are often out of the direct control of users. In sensitive contexts, such as healthcare, where privacy takes priority over functionality, approaches that require centralized data lakes containing user data are far from ideal, and may result in severe limitations in what kinds of models can be developed and what applications can be served.

Other issues that result in privacy concerns are more intimately connected with the mathematical framework of machine learning approaches and, in particular, DL methods. It has been shown that DL models tend to memorize parts of the training data and, potentially, sensitive information within their parameters. Recent research is actively seeking ways to reduce issues caused by this phenomenon. Even though these topics extend beyond distributed and collaborative learning methods, they are still intimately connected to them.

The first MICCAI workshop on Distributed and Collaborative Learning (DCL 2020) aimed at creating a scientific discussion focusing on the comparison, evaluation, and discussion of methodological advancement and practical ideas about machine learning applied to problems where data cannot be stored in centralized databases; where information privacy is a priority; where it is necessary to deliver strong guarantees on the amount and nature of private information that may be revealed by the model as a result of training; and where it's necessary to orchestrate, manage, and direct clusters of nodes participating in the same learning task.

During the first edition of DCL, 12 papers were submitted for consideration and, after peer review, 4 full papers were accepted for presentation and 5 full papers were conditionally accepted, pending improvements and rectifications by authors. Of the conditionally accepted papers, only 4 out of 5 were finally accepted for presentation. Each paper was rigorously reviewed by three reviewers in a double-blind review process. The papers were automatically assigned to reviewers, taking into account and avoiding potential conflicts of interest and recent work collaborations between peers. Reviewers were selected among the most prominent experts in the field from all over the world.

Once the reviews were obtained, the area chairs formulated final decisions over acceptance, conditional acceptance, or rejection of each manuscript. These decisions were always taken according to the reviews and could not be appealed. In case of conditional acceptance, authors had to make substantial changes and improvements to their paper according reviewer feedback. The nature of these changes aimed to increase the scientific validity as well as the clarity of the manuscripts.

Additionally, the workshop organization granted the Best Paper Award to the best submission presented at DCL 2020. The Best Paper Award was assigned as a result of a secret voting procedure where each member of the committee indicated 2 papers worthy of consideration for the award. The paper collecting the majority of votes was then chosen by the committee.

The double-blind review process with three independent reviewers selected for each paper, united with the mechanism of conditional acceptance, as well as the selection and decision process through meta-reviewers, ensured the scientific validity and the high quality of the works presented at the first edition of DCL, making our contribution very valuable to the MICCAI community, and in particular to researchers working on distributed and collaborative learning.

We would therefore like to thank the authors for their contributions, and the reviewers for their dedication and fairness when judging the works of their peers.

August 2020

Shadi Albarqouni
Spyridon Bakas
M. Jorge Cardoso
Bennett Landman
Wenqi Li
Fausto Milletari
Nicola Rieke
Holger Roth
Daguang Xu

Organization

Organization Committee DART 2020

Shadi Albarqouni	Technical University Munich, Germany
M. Jorge Cardoso	King's College London, UK
Konstantinos Kamnitsas	Imperial College London, UK
Fausto Milletari	Verb Surgical, Germany
Nicola Rieke	NVIDIA GmbH, Germany
Daguang Xu	NVIDIA Corporation, USA
Ziyue Xu	NVIDIA Corporation, USA

Program Committee DART 2020

Shekoofeh Azizi	Google, Canada
Ulas Bagci	University of Central Florida, USA
Wenjia Bai	Imperial College London, UK
Gustav Bredell	ETH Zurich, Switzerland
Krishna Chaitanya	ETH Zurich, Switzerland
Cheng Chen	Chinese University of Hong Kong, China
Xiaoran Chen	ETH Zurich, Switzerland
Reuben Dorent	King's College London, UK
Qi Dou	Imperial College London, UK
Ertunc Erdil	ETH Zurich, Switzerland
Enzo Ferrante	Universidad Nacional del Litoral, Argentina
Abhijit Guha Roy	Google, USA
Amelia Jiménez-Sánchez	Pompeu Fabra University, Spain
Maxime Lafarge	Eindhoven University of Technology, The Netherlands
Xiaoxiao Li	Chinese University of Hong Kong, China
Yilin Liu	University of Wisconsin-Madison, USA
Ilja Manakov	Ludwig Maximilian University of Munich, Germany
Nina Miolane	Stanford University, USA
Henry M. Orbes Arteaga	King's College London, UK
Raphael Prevost	ImFusion GmbH, Germany
Chen Qin	Imperial College London, UK
Tobias Ross	German Cancer Research Center, Germany
Mhd Hasan Sarhan	Technical University of Munich, Germany
Hoo-Chang Shin	NVIDIA Corporation, USA
Carole Sudre	King's College London, UK
Gabriele Valvano	IMT Lucca, Italy
Thomas Varsavsky	University College London, UK
Anna Volokitin	ETH Zurich, Switzerland
Yong Xia	Northwestern Polytechnical University, China

Yan Xu   Beihang University, China
Oliver Zettinig   ImFusion GmbH, Germany
Yue Zhang   Case Western Reserve University, USA
Yuyin Zhou   Johns Hopkins University, USA
Xiao-Yun Zhou   Imperial College London, UK

Organization Committee DCL 2020

Shadi Albarqouni   Technical University Munich, Germany
Spyridon Bakas   University of Pennsylvania, USA
M Jorge Cardoso   King's College London, UK
Bennett Landman   Vanderbilt University, USA
Wenqi Li   NVIDIA Ltd, UK
Fausto Milletari   Verb Surgical, Germany
Nicola Rieke   NVIDIA GmbH, Germany
Holger Roth   NVIDIA Corporation, USA
Daguang Xu   NVIDIA Corporation, USA

Program Committee DCL 2020

Amir Alansary   Imperial College London, UK
Reuben Dorent   King's College London, UK
Mark Graham   King's College London, UK
Jonny Hancox   NVIDIA Ltd, USA
Yuankai Huo   Vanderbilt University, USA
Klaus Kades   German Cancer Research Center, Germany
Jayashree Kalpathy-Cramer Harvard Medical School, USA
Sarthak Pati   University of Pennsylvania, USA
G. Anthony Reina   Intel Corporation, USA
Daniel Rubin   Stanford University, USA
Wojciech Samek   Fraunhofer HHI, Germany
Micah J Sheller   Intel Corporation, USA
Ziyue Xu   NVIDIA Corporation, USA
Dong Yang   NVIDIA Corporation, USA
Maximilian Zenk   German Cancer Research Center, Germany

Contents

DCL 2020

DART 2020

α-UNet++: A Data-Driven Neural Network Architecture for Medical Image Segmentation

Yaxin Chen[1,2,3], Benteng Ma[2], and Yong Xia[1,2(✉)]

[1] Research and Development Institute of Northwestern Polytechnical University in Shenzhen, Shenzhen 518057, China
yxia@nwpu.edu.cn
[2] National Engineering Laboratory for Integrated Aero-Space-Ground-Ocean Big Data Application Technology, School of Computer Science and Engineering, Northwestern Polytechnical University, Xi'an 710072, China
[3] Institute of Medical Research, Northwestern Polytechnical University, Xi'an 710072, China

Abstract. UNet++, an encoder-decoder architecture constructed based on the famous UNet, has achieved state-of-the-art results on many medical image segmentation tasks. Despite improved performance, UNet++ introduces densely connected decoding blocks, some of which, however, are redundant for a specific task. In this paper, we propose α-UNet++ that allows us to automatically identify and discard redundant decoding blocks without the loss of precision. To this end, we design an auxiliary indicator function layer to compress the network architecture via removing a decoding block, in which all individual responses are less than a given threshold α. We evaluated the segmentation architecture obtained respectively for liver segmentation and nuclei segmentation, denoted by UNet++^{C}, against UNet and UNet++. Comparing to UNet++, our UNet++^{C} reduces the parameters by 18.89% in liver segmentation and 34.17% in nuclei segmentation, yielding an average improvement of IoU by 0.27% and 0.11% on two tasks. Our results suggest that the UNet++^{C} produced by the proposed α-UNet++ not only improves the segmentation accuracy slightly but also reduces the model complexity considerably.

Keywords: Medical image segmentation · UNet++ · Network compression · Auxiliary indicator function

1 Introduction

With the advances in deep learning, the encoder-decoder networks like UNet [1] and VNet [2] have achieved state-of-the-art performance on many image segmentation tasks [3–5]. They are implemented as one stream from an input to a result with subsampling operations in the encoder and upsampling operations

© Springer Nature Switzerland AG 2020
S. Albarqouni et al. (Eds.): DART 2020/DCL 2020, LNCS 12444, pp. 3–12, 2020.
https://doi.org/10.1007/978-3-030-60548-3_1

in the decoder. The subsampling operations reduce the feature size and enlarge the receptive field, while the upsampling operations increase the feature size back to the original one. The success of these models is largely owing to the skip connections, which collect the shallow and low-level features from encoder and distribute them to those deep and high-level features that share the same resolution in decoder. Other networks like DenseNet [7], H-DenseUNet [8] also show the importance of skip connections for image segmentation [9]. Although transferring the fine-grained features from each encoder layer to the corresponding decoder layer contributes to the dense prediction in segmentation, these skip connections still suffer from the efficient aggregation of dissimilar features due to the gap between low-level and high-level features. The feature mapping in these models faces an inherent tension between semantics and localization: high-level features resolves what the organs or tissues are while low-level features resolves where these objects locate.

To address this issue, UNet++ [6] is proposed to nest and densely connect the decoders at both the same and different scales via convolutional layers and skip connections. The convolutional operation is applied to generate the middle level feature to alleviate the feature gap issue caused by the simple and direct skip connections in UNet. By allowing the aggregation operation to fuse the features of low-level, middle-level, and high-level at each decoder node, UNet++ is powered in feature representation for medical image segmentation. However, the improvement of UNet++ over the classical UNet architecture can be ascribed to the advantages introduced by the extended decoders with many redundant convolutional layers to aggregate the semantic features at multiple levels. Further, the aggregation of semantic features in decoder layers should be more flexible and even data-driven on various datasets or tasks. Moreover, the manually-designed UNet++ may be a suboptimal solution to the fusion of multi-scale and multi-level semantic features on a specific segmentation task.

Recent years have witnessed the success of filter pruning in reducing the inference resource requirements with minimal performance degradation. Li et al. [10] first showed that deep convolutional neural networks (DCNNs) have unimportant filters and then adopted the L_1-norm criterion to eliminate unimportant filters. Zhuo et al. [11] used the spectral clustering to determine which filters are invalid. Suau et al. [12] employed the principal component analysis (PCA) to exploit the correlation among filter responses within the network layer and thus recommended a smaller network that maintains the accuracy as much as possible. Yu et al. [13] proposed ϵ-ResNet that can automatically discard redundant layers, which produce the responses that are smaller than a threshold ϵ, with a marginal or no loss in performance. Huang et al. [14] introduced scaling factors to scale the outputs of specific structures, such as neurons, groups or residual blocks, and thus pruned the unimportant structures by forcing the corresponding factors to zero.

In this paper, we propose α-UNet++, a data-driven pruning method that allows us to automatically skip redundant decoding blocks in UNet++. With adaptive diversity and more pertinent features, α-UNet++ can efficiently allevi-

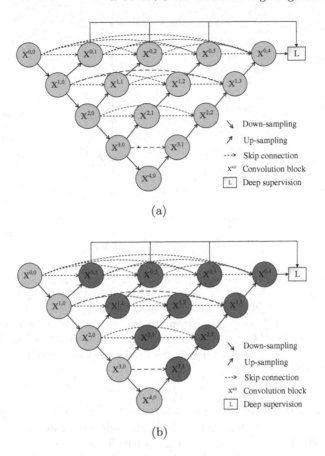

(a)

(b)

Fig. 1. Architectures of (a) UNet++ and (b) α-UNet++. UNet++ is constructed from U-Net of varying depths by sharing the same encoder. Each node in the UNet++ decoder fuses the features of all its nodes on the same horizontal line. Deep supervision is introduced to UNet++ by adding the loss function to each of the four semantic levels. α-UNet++ introduces an auxiliary indicator function to the decoder nodes of UNet++, highlighted in blue circles, which can select important decoding blocks and discard redundant ones. (Color figure online)

ate the feature gap issue in UNet++. We evaluated α-UNet++ on two medical image segmentation tasks (*i.e.*, the liver segmentation and nuclei segmentation). Our results suggest that α-UNet++ can automatically prune the structure to adapt to different segmentation tasks. The network it generated has greatly reduced parameters and slightly improved segmentation performance.

2   Datasets

Two datasets were used for this study.

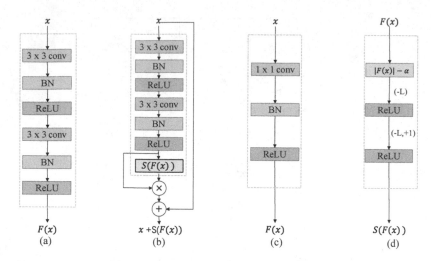

Fig. 2. Structures of four types of convolutional blocks: (a) a convolutional block in standard UNet or UNet++, (b) a decoding block in α-UNet++, (c) a compressed convolutional block in UNet++^{C}, and (d) an auxiliary indicator function S(F(X)) that uses two sequential ReLU layers. Comparing to (a), an auxiliary indicator and residual connection are introduced to (b), aiming to discard redundant decoding blocks. In the structure of S(F(X)), L is a large positive constant and α is a threshold that controls the compression ratio of the model.

The liver segmentation dataset contains 131 labeled CT scans, provided by the Medical Segmentation Decathlon (MSD) Challenge[1]. Each scan is equipped with a manually annotated ground truth, in which there are two types of labels: the liver and tumors. For this study, we only consider whether a voxel belongs to the liver or not. We randomly split this dataset into a training set of 100 scans, a validation set of 15 scans, and a test set of 16 scans.

The nuclei segmentation dataset contains 670 segmented nuclei images, provided by the 2018 Data Science Bowl Segmentation Challenge[2]. The images were acquired under a variety of conditions and vary in the cell type, magnification, and imaging modality (brightfield vs. fluorescence). Each image is equipped with a human-annotated segmentation mask of each nucleus. Masks are not allowed to overlap (*i.e.*, no pixel belongs to two masks). We randomly split this dataset into a training set of 336 images, a validation set of 134 images, and a test set of 200 images.

3   Method

Given a medical image segmentation task, our goal is to design a method that can selectively adjust the strategy of fusing multi-level features in UNet++ in a

---

[1] https://decathlon-10.grand-challenge.org.
[2] https://www.kaggle.com/c/data-science-bowl-2018/data.

data-dependent fashion (see Fig. 1). To this end, we first introduce a criterion to identify important decoding blocks, and then replace each redundant decoding block that include multiple convolutional layers with a modified skip connection.

3.1   Pruning Decoding Blocks

For each decoding block in UNet++, the output y is the sum of the identity mapping and residual mapping of the input feature map x, shows as follows [15]

$$y = x + F(x),\tag{1}$$

where $F(x)$ is the output of the convolutional layers in the decoding block.

To measure the importance of each decoding block, we introduce an auxiliary indicator function $S(F(x))$, which can restraint the redundant block if all element responses in $F(x)$ are less than a threshold α. Thus, we define the indicator function as

$$S(F(x)) = \begin{cases} 0, & |F(x)^i| < \alpha, i \in \{0, 1, ..., n-1) \\ 1, & otherwise, \end{cases}\tag{2}$$

where $F(x)^i$ indicates the i-th element in $F(x)$. The indicator function is implemented as an embedding block with two modified sparse activation layers with hyperparameters α and L. The structure of $S(F(x))$ is shown in Fig. 2(d) and we can represent the function layer as follow:

$$S(F(x)) = ReLU\left((ReLU\left((|F(x)| - \alpha) \times (-L)\right)) \times (-L) + 1\right)\tag{3}$$

$$ReLU\,(x) = \begin{cases} x, & x > 0 \\ 0, & otherwise \end{cases}\tag{4}$$

Here, L is a large positive constant and α is a threshold that controls the compression ratio of the model. Two sequential ReLU layers are utilized to convert the output to 0 or 1. In this case, the indicator function layers can select important decoding blocks by setting the output to 1 and discard redundant blocks by changing the output to 0. When the indicator layer removes a redundant block during the pruning process, we replace this block with a residual connection to ensure the continuity of feature transmission and gradient back propagation, which is vital for the dense prediction. Figure 2(b) shows the structure of a decoding block in α-UNet++.

Note that we only select the optimal operations of feature transmission among decoders to learn the optimal routing to make full use of the feature generated in the encoder. In addition, we believe that the compression of encoder will greatly damage the model performance and hence maintain the encoder structure during the selection of decoding blocks.

3.2   UNet++^{C}

After applying α-UNet++ to a specific medical image segmentation task, we can have a sub-optimal decoder architecture for UNet++. Then, we compress UNet++ via replacing each redundant decoding block with a convolutional layer with the 1×1 kernels, which adjusts the number of feature channels to facilitate the aggregation of multi-level features (see Fig. 2(c)). The resultant compressed network is denoted by UNet++^{C}.

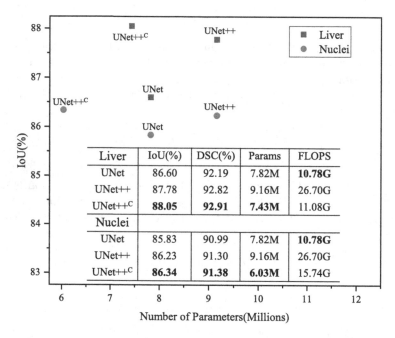

The table embedded within the figure:

Liver	IoU(%)	DSC(%)	Params	FLOPS
UNet	86.60	92.19	7.82M	**10.78G**
UNet++	87.78	92.82	9.16M	26.70G
UNet++^{C}	**88.05**	**92.91**	**7.43M**	11.08G
Nuclei				
UNet	85.83	90.99	7.82M	**10.78G**
UNet++	86.23	91.30	9.16M	26.70G
UNet++^{C}	**86.34**	**91.38**	**6.03M**	15.74G

Fig. 3. Segmentation results (IoU: % and DSC: %) for U-Net, UNet++ and UNet++^{C} for different datasets.

3.3   Implementation and Evaluation

The proposed α-UNet++ was optimized by using the Adam optimizer with a learning rate of 3e-4. The early-stop mechanism is utilized to avoid over-fitting.

The obtained UNet++^{C} was evaluated against UNet and UNet++ on the liver segmentation dataset and nuclei segmentation dataset. We chose those two as competing models since they work well on many medical image segmentation tasks and provide the basis for our model (see Fig. 2). To make a fair comparison, all convolutional blocks $X^{i,j}$ use k kernels of size 3×3 where $k = 32 \times 2^i$. UNet++ appends a 1×1 convolutional layer followed by the Sigmoid activation function to the outputs of blocks $\{X^{0,j}, j \in 1, 2, 3, 4\}\}$ to introduce deep supervision.

All experiments were conducted based on the Pytorch framework on a desktop with a NVIDIA TITAN X (Pascal) 12 GB GPUs. The segmentation results were evaluated by the Intersection over Union (IoU) [6] and Dice Similarity Coefficient (DSC) [8].

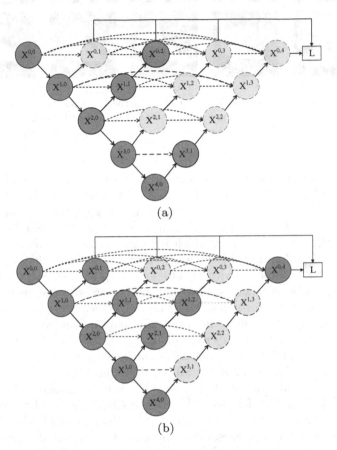

(a)

(b)

Fig. 4. Architectures of the obtained UNet++^{C} for (a) the liver segmentation task and (b) nuclei segmentation task, in which seven and five decoding blocks are discarded, respectively. The discarded decoding blocks are highlighted in gray and dotted circles. (Color figure online)

4 Results and Discussion

Result. The segmentation network (*i.e.*, UNet++^{C}) automatically generated by our α-UNet++ for the liver segmentation task and nuclei segmentation task are displayed in Fig. 4, where the convolutional blocks shown in gray and dotted circles are discarded.

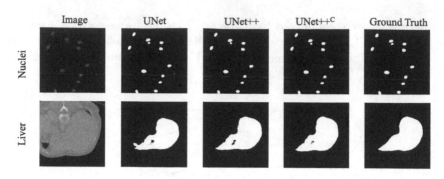

Fig. 5. Comparison of segmentation results: (top row) a nuclei image and (bottom row) a liver slice, the segmentation results obtained by (second column) UNet, (third column) UNet++, and (fourth column) UNet++^{C}, and (right column) ground truth.

Figure 3 shows the IoU and DSC values of the segmentation results obtained by applying U-Net, UNet++ and UNet++^{C} to the liver segmentation dataset and nuclei segmentation dataset, respectively. It also gives the number of parameters and FLOPS of each segmentation network. It reveals that (1) UNet++ outperforms UNet on both tasks, which is largely attributed to the nested and densely connected decoding blocks at both the same and different scales in UNet++; and (2) compared with UNet++, UNet++^{C} achieves a substantial performance gain in terms of both IoU and DCS even with the greatly reduced number of parameters and computational cost. Specifically, on the liver segmentation task, UNet++^{C} reduces the number of parameters by 18.89% and the FLOPS by 58.50%, and improves the IoU by 0.27% and DSC by 0.09%. On the nuclei segmentation, UNet++^{C} reduces the number of parameters by 34.17% and the FLOPS by 41.05%, but improves IoU by 0.11% and DSC by 0.08%. Figure 5 visualizes the segmentation results obtained on an nuclei image and a liver slice using U-Net, UNet++ and UNet++^{C}. It shows that the results produced by UNet++^{C} are more similar to the ground truth than those of other segmentation models. Therefore, it suggests that the UNet++^{C} automated generated by our α-UNet++ can produce slightly more accurate segmentation results than UNet++ but using significantly reduced parameters and computational cost.

Table 1. Average time cost of using UNet, UNet++, or UNet++^{C} to segment a liver slice or a nuclei image. Experiments were performed on a desktop with one NVIDIA TITAN X (Pascal) GPU.

	Inference time (ms)		
	UNet	UNet++	UNet++^{C}
Liver	**52**	77	64
Nuclei	**202**	317	276

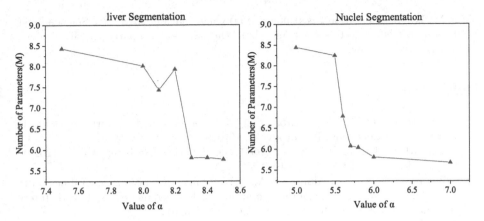

Fig. 6. Plot of the number of parameters in UNet++^{C} versus the value of α on two segmentation tasks.

Value of α. The hyperparameter α controls the compression ratio of the model. To discuss the setting of this hyperparameter, we repeated our segmentation experiments using different values of α and plotted the number of model parameters versus the value of α in Fig. 6. It shows that a large α leads to a higher compression ration. Considering the trade-off between the accuracy and complexity of UNet++^{C}, we set α to 8.1 for the liver segmentation task and to 5.8 for the nuclei segmentation task.

Inference Time. Tabel 1 shows the average time cost of using UNet, UNet++, or UNet++^{C} to segment a liver slice or a nuclei image. It shows that UNet++^{C} can segment a liver slice within a second and segment a nuclei image within three seconds, which is somewhat slower than UNet, but faster than UNet++.

5   Conclusion

In this paper, we propose α-UNet++ that can automatically discard redundant decoding blocks to address the need for lightweight networks in medical image segmentation. It uses an auxiliary indicator function to select the redundant blocks in the network. If all the individual responses from a block are smaller than a threshold α, the block is treated as redundant and will be discarded. We evaluated the resultant UNet++^{C} on two tasks: liver and nuclei segmentation. Our results suggest that our UNet++^{C} not only improves the segmentation accuracy slightly but also reduces model parameters and FLOPS considerably.

Acknowledgment. This work was supported in part by the Science and Technology Innovation Committee of Shenzhen Municipality, China, under Grants JCYJ20180306171334997, and in part by the National Natural Science Foundation of China under Grants 61771397. We appreciate the efforts devoted by the organizers of the Medical Segmentation Decathlon (MSD) Challenge and 2018 Data Science Bowl Segmentation Challenge to collect and share the data for comparing medical image segmentation algorithms.

References

1. Ronneberger, O., Fischer, P., Brox, T.: U-Net: convolutional networks for biomedical image segmentation. In: Navab, N., Hornegger, J., Wells, W.M., Frangi, A.F. (eds.) MICCAI 2015. LNCS, vol. 9351, pp. 234–241. Springer, Cham (2015). https://doi.org/10.1007/978-3-319-24574-4_28
2. Milletari, F., Navab, N., Ahmadi, S.A.: V-Net: fully convolutional neural networks for volumetric medical image segmentation. In: Proceedings of the Fourth International Conference on 3D Vision, pp. 565–571 (2016)
3. Shen, D., Wu, G., Suk, H.I.: Deep learning in medical image analysis. Ann. Rev. Biomed. Eng. **19**, 221–248 (2017)
4. Tajbakhsh, N., Jeyaseelan, L., Li, Q., Chiang, J., Wu, Z., Ding, X.: Embracing imperfect datasets: a review of deep learning solutions for medical image segmentation. arXiv preprint arXiv:1908.10454 (2019)
5. Litjens, G., et al.: A survey on deep learning in medical image analysis. Med. Image Anal. **42**, 60–88 (2017)
6. Zhou, Z., Siddiquee, M.M.R., Tajbakhsh, N., Liang, J.: UNet++: redesigning skip connections to exploit multiscale features in image segmentation. IEEE Trans. Med. Imaging **39**(6), 1856–1867 (2019)
7. Huang, G., Liu, Z., Van Der Maaten, L., Weinberger, K.Q.: Densely connected convolutional networks. In: Proceedings of the IEEE Conference on Computer Vision and Pattern Recognition, pp. 4700–4708 (2017)
8. Li, X., Chen, H., Qi, X., Dou, Q., Fu, C.W., Heng, P.A.: H-DenseUNet: hybrid densely connected UNet for liver and tumor segmentation from CT volumes. IEEE Trans. Med. Imaging **37**(12), 2663–2674 (2018)
9. Drozdzal, M., Vorontsov, E., Chartrand, G., Kadoury, S., Pal, C.: The importance of skip connections in biomedical image segmentation. In: Carneiro, G., et al. (eds.) LABELS/DLMIA -2016. LNCS, vol. 10008, pp. 179–187. Springer, Cham (2016). https://doi.org/10.1007/978-3-319-46976-8_19
10. Li, H., Kadav, A., Durdanovic, I., Samet, H., Graf, H.P.: Pruning filters for efficient convnets. arXiv preprint arXiv:1608.08710 (2016)
11. Zhuo, H., Qian, X., Fu, Y., Yang, H., Xue, X.: SCSP: spectral clustering filter pruning with soft self-adaption manners. arXiv preprint arXiv:1806.05320 (2018)
12. Suau, X., Zappella, L., Palakkode, V., Apostoloff, N.: Principal filter analysis for guided network compression. arXiv preprint arXiv:1807.10585 (2018)
13. Yu, X., Yu, Z., Ramalingam, S.: Learning strict identity mappings in deep residual networks. In: Proceedings of the IEEE Conference on Computer Vision and Pattern Recognition, pp. 4432–4440 (2018)
14. Huang, Z., Wang, N.: Data-driven sparse structure selection for deep neural networks. In: Proceedings of the IEEE Conference on European Conference on Computer Vision, pp. 304–320 (2018)
15. He, K., Zhang, X., Ren, S., Sun, J.: Deep residual learning for image recognition. In: Proceedings of the IEEE Conference on Computer Vision and Pattern Recognition, pp. 770–778 (2016)

DAPR-Net: Domain Adaptive Predicting-Refinement Network for Retinal Vessel Segmentation

Zichun Huang[1], Hongyan Mao[1(✉)], Ningkang Jiang[1], and Xiaoling Wang[2]

[1] Shanghai Key Laboratory of Trustworthy Computing,
East China Normal University, Shanghai, China
{hymao,nkjiang}@sei.ecnu.edu.cn
[2] Computer Science Department, Shanghai Lixin University of Accounting
and Finance, Shanghai, China

Abstract. The convolutional neural networks (CNN) have been used in various medical image segmentation tasks. However, the training of CNN extremely relies on large amounts of sample images and precisely annotated labels, which is difficult to get in medical field. Domain adaptation can utilize limited labeled images of source domain to improve the performance of target domain. In this paper, we propose a novel domain adaptive predicting-refinement network called DAPR-Net to perform domain adaptive segmentation task on retinal vessel images. In order to mitigate the gap between two domains, the Contrast Limited Adaptive Histogram Equalization (CLAHE) is employed in the preprocessing operations. Since the segmentation result generated by only predicting module can be affected by domain shift, refinement module is used to produce more precise segmentation results and further reduce the harmful impact of domain shift. Atrous convolution is also adopted in both predicting module and refinement module to capture wider and deeper semantic features. Our method has advantages over previous works based on adversarial networks, because in our method smoothing domain shift with preprocessing has little overhead and the data from target domain is not needed when training. Experiments on different retinal vessel datasets demonstrate that the proposed method improves accuracy of segmentation results in dealing with domain shift.

1 Introduction

Over the past two decades, convolutional neural networks (CNN) has achieved remarkable success in various fields. However, training a well-performing neural network requires a large amount of well-sampled data and precisely annotated labels, which is scarcely available in the medical field. On one hand, due to the acquisition device noise, imaging angles and resolutions, image acquisition techniques and other reasons [1], different datasets may differ in resolution, brightness, color, etc. even on the same subject, which is called domain shift or domain gap. Since annotating images is very expensive and time-consuming,

© Springer Nature Switzerland AG 2020
S. Albarqouni et al. (Eds.): DART 2020/DCL 2020, LNCS 12444, pp. 13–22, 2020.
https://doi.org/10.1007/978-3-030-60548-3_2

people want to utilize labeled data in one or more relevant source domains to execute new tasks in a target domain [2], which is so-called domain adaptation. In other words, domain adaptation is used to solve domain shift caused by the inconsistency between source domain and target domain.

Performing domain adaptive segmentation task on retinal vessel images is a challenging task, because fundus images from different datasets are varying in terms of size, angle and resolution and retinal vessel segmentation requires processing models with very high resolution. Though some methods have been proposed [1,3,4], the significance of preprocessing is ignored when original RGB images are used as the input of their neural networks. And CLAHE has been proved effective in medical tasks [5]. Compared with the unsupervised domain adaptation (UDA) methods [1,4] where unlabeled images of target domain are needed for training, we further relax the data requirements of our method that only labeled images of source domain are needed for training. Meanwhile, our method avoids the disadvantages brought by adversarial training, such as large training overhead and difficulty in convergence.

In this paper, we propose a novel network called domain adaptive predicting-refinement network (DAPR-Net) to perform domain adaptive segmentation task on retinal vessel images. Specifically, in addition to the traditional preprocessing operations, including padding, flipping and rotating, CLAHE with a relatively large clip limit is used to mitigate the gap between domains. And our proposed network is mainly composed of a predicting module and a refinement module with atrous convolution blocks. Predicting module can generate rough predicted results by taking CLAHE enhanced fundus images as input, which still can be affected by domain shift. Refinement module learns to use prior knowledge of vessel morphology to refine the intermediate results by taking the predicted results of predicting module alone as input. Thus, refinement module can produce more precise segmentation results and further reduce the harmful impact of domain shift. Atrous convolution [6] block, which is used in both predicting module and refinement module, can guide the network to capture wider and deeper semantic features [7]. When training our network, random noise is added to the intermediate results of predicting module to improve the generalization ability of refinement module. Experiments on five different public retinal vessel datasets demonstrate that the proposed method improves the accuracy of segmentation results in dealing with domain shift.

The main contributions of this work are summarized as: (1) Reasonable and adaptable preprocessing operations are given that can significantly mitigate domain gap. (2) A novel predicting-refinement network architecture is proposed to tackle the harmful effect of domain shift. (3) Experiments are taken on different public datasets to prove the effectiveness of the proposed method.

2   Related Works

Domain adaptation is a particular case of transfer learning, and many approaches have been proposed to solve domain shift in image segmentation. One category

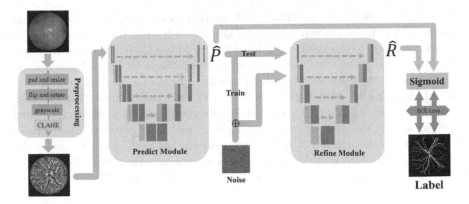

Fig. 1. The overall architecture of proposed DAPR-Net.

of the mainstream methods is unsupervised domain adaptation (UDA) [2], but representative images of target domain are scarcely available in many cases [3]. And adversarial-based methods are often used [2] in UDA. Liu et al. [12] and Dong et al. [9] used adversarial learning to capture domain-invariant high-level features. Javanmardi et al. [4] used adversarial learning to get domain-invariant predicted results. However, adversarial-based methods are hard to train and less effective possibly due to the fact that high-dimensional feature maps used for segmentation are more challenging to align compared to the features used for classification [18]. Chen et al. [8] used GAN to generate domain-invariant stylized images and then performed segmentation tasks on these images, but GAN may introduce some unknown changes to the image, causing some details to be changed or get lost and thus resulting in a decrease in the interpretability of the model. Tajbakhsh et al. [3] used an error injection module and an error prediction module to inject error patterns and correct the injected error respectively, but the error injection module often fails to generate satisfactory results. In addition, apart from the medical field, another field where domain adaptation is very popular is the semantic segmentation of urban scenes, and many sophisticated methods have been proposed [10,15,17,19]. However, since the medical field places more emphasis on the accuracy of segmentation than the accuracy of target detection, directly transforming these methods to medical image segmentation will fail to get satisfactory results.

3 Method

In this section, we give the detailed network architecture of the proposed DAPR-Net for retinal vessel segmentation. The overall architecture of our DAPR-Net is illustrated in Fig. 1 and detailed information can be found in Fig. 2. The motivation of our method is that although the fundus images are different, the morphology of the vessels is similar. We use the learned vessel morphology knowledge to improve the segmentation accuracy after minimizing the domain shift.

The input fundus image firstly will be preprocessed to mitigate the gap between known and unknown domains. Then the output of preprocessing operations will be input to the Predict Module (PM). PM is a U-shape convolution network and is responsible for generating rough predictions $\hat{P}$. Though the domain gap is significantly mitigated by preprocessing operations, fundus images from different domains are still slightly different. Hence, Refinement Module (RM) is responsible for learning to use prior knowledge of vessel morphology, which is similar among different domains, to refine $\hat{P}$ and thus generate more precise and clear prediction results $\hat{R}$. When we train our network, some normally distributed random noise is added to $\hat{P}$ to improve the generalization ability of the RM. When we test our network, $\hat{P}$ is directly passed to the RM.

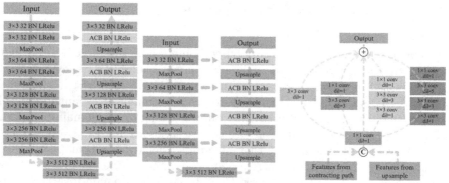

(a) Predicting Module   (b) Refinement Module   (c) Atrous Convolution Block

Fig. 2. The detailed architecture of the predicting module, refinement module and atrous convolution block of our network. In the figure, the '33 32 BN LRelu', for example, means that the convolution layer is composed of a 33 convolution layer with filters of 32 and then is followed by a batch normalization layer and a Leaky-Relu activation function.

PM and RM can be trained jointly. Our loss is defined as the Binary Cross Entropy (BCE) loss of both PM and RM:

$$\mathcal{L} = \ell_{bce}(\hat{P}, y) + \ell_{bce}(\hat{R}, y) \tag{1}$$

$$\ell_{bce}(\hat{y}, y) = -\frac{1}{n}\Sigma_i(\hat{y}[i] * \log(y[i]) + (1 - \hat{y}[i]) * \log(1 - y[i])) \tag{2}$$

where y is the ground truth label and n is the batch size.

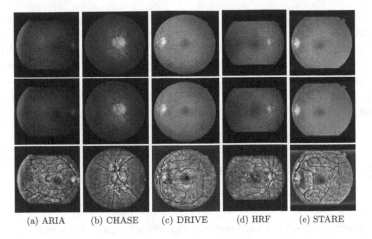

(a) ARIA   (b) CHASE   (c) DRIVE   (d) HRF   (e) STARE

Fig. 3. Sample images from different datasets and the results after preprocessing. The lines from top to bottom are the original image, the grayscale image and the image enhanced by CLAHE.

3.1 Preprocessing Operations

Before training, we first preprocess the images to enhance them and expand the dataset. To be specific, we perform traditional basic preprocessing operations, including resizing them to 512512, flipping and then rotating them every 15°. Then, grayscale transformation and CLAHE with the clip limit of 12.0 is performed sequentially. A relatively large clip limit is adopted, making the domain gap of different domains as small as possible. Sample images from different datasets and the results after preprocessing are shown in Fig. 3.

3.2 Network Architecture

Predicting Module (PM). Unet [14] is used as the backbone of our network. In U-shape Networks architecture, each block of decoder contains an upsample layer and two convolution layers. In the proposed method, we replace the first convolution layer with Atrous Convolution Block (ACB). Each of these convolution layers is followed by a batch normalization and a Leaky-Relu activation function.

Refinement Module (RM). Refinement module is mainly inspired by the residual refinement module in BASNet [13]. We also replace the 33 convolution layer in the decoder with ACB. Like PM, each of these convolution layers is followed by a batch normalization and a Leaky-Relu activation function.

Atrous Convolution Block (ACB). Atrous convolution block is mainly inspired by [6] and [7]. Atrous convolution is proved to be effective because it can enlarge the field of view of lters to incorporate multi-scale context [6]. With larger field of view, wider and deeper semantic features can be captured [7], helping our model to distinguish between noise and the vessel. When the features from the contracting path and the upsample layer are passed to the ACB, we first concatenate them and use a 11 convolution to reduce the number of filters by half. Then atrous convolution of four different levels is used. Finally, the output features of atrous convolution and the original features are added directly as the output of ACB.

4   Experiments and Results

4.1   Datasets

We took experiments on five different public retinal vessel segmentation datasets: ARIA, CHASE, DRIVE, HRF, STARE. The selected datasets vary in terms of size, angle and resolution, acquisition device and acquisition techniques, allowing us to evaluate the effectiveness of our method under different situations. Their specific resolution, vessel ratio and partition information are shown in Table 1, where the vessel ratio means the proportion of vessels reflected by the label in the entire picture.

Table 1. Specific resolution, vessel ratio and partition information of different datasets.

Dataset	Resolutions	Vessel ratio	Data splits		
			Train	Val	Test
ARIA	576768	8.115%	–	–	143
CHASE	960999	6.934%	20	2	6
DRIVE	584565	8.695%	18	2	20
HRF	23363504	7.709%	–	–	45
STARE	605700	7.604%	–	–	20

4.2   Baselines and Evaluation Scenarios

DAPR-Net is compared against Unet [14], AdaptSegNet [17], V-GAN [16], Javanmardi et al. [4] and ErrorNet [3]. We implement Unet with Leaky-Relu activation function following the practice suggested in [11]. We implement AdaptSegnet, one of the most sophisticated networks in semantic segmentation of urban scenes, because we want to prove that even the most sophisticated method in another field cannot perform well in medical image segmentation task. AdaptSegNet [17] and Javanmardi et al. [4] are UDA methods. Unet [14], V-GAN [16] and ErrorNet [3], like ours, use only labeled images from source domain for training and no images from target domain are needed.

Based on the data partition shown in Table 1, we train all the methods above on DRIVE and CHASE respectively and then choose the model performing best on validation datasets for test. DRIVE and CHASE are the datasets with the highest and lowest vessel ratio, which can reflect the generalization ability of each method above in two different situations.

AUC (Area Under Curve) and Dice are used for comparison. Dice is a common indicator in medical images.

$$Dice = \frac{2TP}{2TP + FP + FN} \tag{3}$$

where TP, FP and FN represent the number of true positives, false positives and false negatives respectively.

4.3   Training Strategy

Our proposal is implemented on Pytorch framework. We utilize the Adam optimizer to train our network and the learning rate is set 0.0015. The batch size is 3. The noise mentioned above follows a normal distribution with a mean of 0 and a variance of 0.1. The rest hyper-parameters are set to the default values.

4.4   Ablation Study

Since that previous works [1,3,4] use RGB images as the input of their networks and our proposed preprocessing operations can also be applied in various methods, we compare all the methods above with and without CLAHE in preprocessing respectively. Also, we investigate the effectiveness of each part of DAPR-Net by implementing our proposed DAPR-Net with PM only and then progressively extending it with RM and noise.

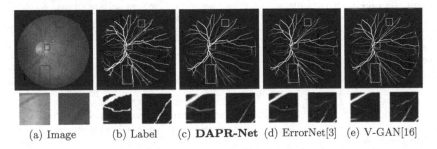

(a) Image   (b) Label   (c) **DAPR-Net** (d) ErrorNet[3] (e) V-GAN[16]

Fig. 4. Visualization of an image from DRIVE and the models are trained on CHASE.

4.5   Results

The dice results of each method under the evaluation scenarios are summarized in Table 2. As we can see, the proposed DAPR-Net outperforms all the segmentation methods above. In particular, our proposal achieves an average dice of 0.7772 and 0.7703 when training on DRIVE dataset and CHASE dataset. It can also be seen that methods with CLAHE perform better than that without CLAHE in most cases, hence, our proposed preprocessing operations are adaptable. However, even with the help of CLAHE, the previous methods still have a certain gap in performance compared to ours. What's more, compared with the implementation without refinement module and noise, models with all parts trained jointly gain a performance increase of 0.0093 and 0.0077 when training on DRIVE and CHASE respectively. In addition, the average AUC results are summarized in Table 3 where CLAHE is adopted in all methods. And DAPR-Net also achieves the best results. Figure 4 shows that DAPR-Net produces clearer vessel segmentation results and is more resistant to noise than previous works.

Table 2. Comparison of the proposed method and other methods on five datasets. Dice is used for comparison.

Training on DRIVE

Method	Configurations	ARIA	CHASE	DRIVE	HRF	STARE	Avg.
Unet [14]	+CLAHE	0.7062	0.6328	0.8283	0.7315	0.7916	0.7381
		0.7320	0.7512	0.8260	0.7452	0.7894	0.7687
AdaptSegNet [17]	+CLAHE	0.3668	0.4708	0.5960	0.4758	0.4348	0.4688
		0.5452	0.5198	0.6186	0.4468	0.5619	0.5385
V-GAN [16]	+CLAHE	0.7045	0.6203	0.8224	0.7387	**0.7996**	0.7371
		0.7261	0.7392	0.8223	0.7615	0.7884	0.7675
Javanmardi et al. [4]	+CLAHE	0.6860	0.6562	0.8269	0.7328	0.7643	0.7332
		0.7286	0.7498	0.8268	0.7533	0.7908	0.7698
ErrorNet [3]	+CLAHE	0.7004	0.6754	0.8276	0.7429	0.7806	0.7454
		0.7316	0.7612	0.8274	0.7482	0.7843	0.7705
DAPR-Net (ours)	CLAHE+PM	0.7284	0.7515	0.8279	0.7517	0.7800	0.7679
	CLAHE+PM+RM	0.7304	0.7586	0.8194	**0.7706**	0.7892	0.7736
	CLAHE+PM+RM+noise	**0.7367**	**0.7735**	**0.8280**	0.7526	0.7954	**0.7772**

Training on CHASE

Method	Configurations	ARIA	CHASE	DRIVE	HRF	STARE	Avg.
Unet [14]	+CLAHE	0.6599	0.8107	0.7104	0.7457	0.7505	0.7354
		0.6932	0.8161	0.7467	0.6976	0.7637	0.7434
AdaptSegNet [17]	+CLAHE	0.1944	0.4319	0.5673	0.4640	0.4000	0.4115
		0.5533	0.5761	0.5654	0.4249	0.3073	0.4854
V-GAN [16]	+CLAHE	0.6638	0.8149	0.7189	0.7454	0.7368	0.7360
		0.6990	0.8229	0.7474	0.7490	0.7652	0.7567
Javanmardi et al. [4]	+CLAHE	0.6784	0.8192	0.7469	**0.7718**	0.7731	0.7579
		0.6928	0.8189	0.7341	0.7228	0.7663	0.7470
ErrorNet [3]	+CLAHE	0.6675	0.8208	0.7372	0.7485	0.7628	0.7474
		0.6931	0.8156	0.7556	0.7284	0.7630	0.7511
DAPR-Net (ours)	CLAHE+PM	0.7018	0.8211	0.7637	0.7452	0.7811	0.7626
	CLAHE+PM+RM	0.7042	**0.8234**	0.7663	0.7538	0.7778	0.7651
	CLAHE+PM+RM+noise	**0.7225**	0.8166	**0.7822**	0.7423	**0.7877**	**0.7703**

Table 3. Comparison of different methods (with CLAHE) trained on two datasets with respect to the average AUC calculated on all five datasets.

	Unet	AdaptSegNet	V-GAN	Javanmardi et al.	ErrorNet	DAPR-Net (ours)
DRIVE	0.97977	0.92405	0.97720	0.98042	0.98064	**0.98102**
CHASE	0.97326	0.87771	0.97797	0.97267	0.96880	**0.97959**

5 Conclusion

In this paper, we propose a novel network called domain adaptive predicting-refinement network (DAPR-Net) to perform domain adaptive segmentation task on retinal vessel images. CLAHE is employed in the preprocessing to mitigate the gap between two domains. And the learned prior knowledge of vessel morphology is used to improve the segmentation accuracy in refinement module. Evaluation is executed on five public datasets to prove the effectiveness of each part of our network and the performance gain on the accuracy of segmentation results.

References

1. Zhuang, J., Chen, Z., Zhang, J., Zhang, D., Cai, Z.: Domain adaptation for retinal vessel segmentation using asymmetrical maximum classifier discrepancy. In: Proceedings of the ACM Turing Celebration Conference-China, pp. 1–6, May 2019
2. Wang, M., Deng, W.: Deep visual domain adaptation: a survey. Neurocomputing **312**, 135–153 (2018)
3. Tajbakhsh, N., Lai, B., Ananth, S., Ding, X.: ErrorNet: learning error representations from limited data to improve vascular segmentation. arXiv preprint arXiv:1910.04814 (2019)
4. Javanmardi, M., Tasdizen, T.: Domain adaptation for biomedical image segmentation using adversarial training. In: 2018 IEEE 15th International Symposium on Biomedical Imaging (ISBI 2018), pp. 554–558. IEEE, April 2018
5. Pisano, E.D., et al.: Contrast limited adaptive histogram equalization image processing to improve the detection of simulated spiculations in dense mammograms. J. Digit. Imaging **11**(4), 193 (1998). https://doi.org/10.1007/BF03178082
6. Chen, L.C., Papandreou, G., Schroff, F., Adam, H.: Rethinking atrous convolution for semantic image segmentation. arXiv preprint arXiv:1706.05587 (2017)
7. Gu, Z., et al.: CE-Net: context encoder network for 2D medical image segmentation. IEEE Trans. Med. Imaging **38**(10), 2281–2292 (2019)
8. Chen, C., Dou, Q., Chen, H., Qin, J., Heng, P.A.: Synergistic image and feature adaptation: towards cross-modality domain adaptation for medical image segmentation. In: Proceedings of the AAAI Conference on Artificial Intelligence, vol. 33, pp. 865–872, July 2019
9. Dong, N., Kampffmeyer, M., Liang, X., Wang, Z., Dai, W., Xing, E.: Unsupervised domain adaptation for automatic estimation of cardiothoracic ratio. In: Frangi, A.F., Schnabel, J.A., Davatzikos, C., Alberola-López, C., Fichtinger, G. (eds.) MICCAI 2018. LNCS, vol. 11071, pp. 544–552. Springer, Cham (2018). https://doi.org/10.1007/978-3-030-00934-2_61

10. Hoffman, J., et al.: CyCADA: cycle-consistent adversarial domain adaptation. arXiv preprint arXiv:1711.03213 (2017)
11. Isensee, F., Kickingereder, P., Wick, W., Bendszus, M., Maier-Hein, K.H.: No new-net. In: Crimi, A., Bakas, S., Kuijf, H., Keyvan, F., Reyes, M., van Walsum, T. (eds.) BrainLes 2018. LNCS, vol. 11384, pp. 234–244. Springer, Cham (2019). https://doi.org/10.1007/978-3-030-11726-9_21
12. Liu, P., Kong, B., Li, Z., Zhang, S., Fang, R.: CFEA: collaborative feature ensembling adaptation for domain adaptation in unsupervised optic disc and cup segmentation. In: Shen, D., et al. (eds.) MICCAI 2019. LNCS, vol. 11768, pp. 521–529. Springer, Cham (2019). https://doi.org/10.1007/978-3-030-32254-0_58
13. Qin, X., Zhang, Z., Huang, C., Gao, C., Dehghan, M., Jagersand, M.: BASNet: boundary-aware salient object detection. In: Proceedings of the IEEE Conference on Computer Vision and Pattern Recognition, pp. 7479–7489 (2019)
14. Ronneberger, O., Fischer, P., Brox, T.: U-Net: convolutional networks for biomedical image segmentation. In: Navab, N., Hornegger, J., Wells, W.M., Frangi, A.F. (eds.) MICCAI 2015. LNCS, vol. 9351, pp. 234–241. Springer, Cham (2015). https://doi.org/10.1007/978-3-319-24574-4_28
15. Sankaranarayanan, S., Balaji, Y., Jain, A., Lim, S.N., Chellappa, R.: Unsupervised domain adaptation for semantic segmentation with GANs. arXiv preprint arXiv:1711.06969 (2017)
16. Son, J., Park, S.J., Jung, K.H.: Retinal vessel segmentation in fundoscopic images with generative adversarial networks. arXiv preprint arXiv:1706.09318 (2017)
17. Tsai, Y.H., Hung, W.C., Schulter, S., Sohn, K., Yang, M.H., Chandraker, M.: Learning to adapt structured output space for semantic segmentation. In: Proceedings of the IEEE Conference on Computer Vision and Pattern Recognition, pp. 7472–7481 (2018)
18. Wu, Z., et al.: DCAN: dual channel-wise alignment networks for unsupervised scene adaptation. In: Ferrari, V., Hebert, M., Sminchisescu, C., Weiss, Y. (eds.) ECCV 2018. LNCS, vol. 11209, pp. 535–552. Springer, Cham (2018). https://doi.org/10.1007/978-3-030-01228-1_32
19. Zhang, Y., Qiu, Z., Yao, T., Liu, D., Mei, T.: Fully convolutional adaptation networks for semantic segmentation. In: Proceedings of the IEEE Conference on Computer Vision and Pattern Recognition, pp. 6810–6818 (2018)

Augmented Radiology: Patient-Wise Feature Transfer Model for Glioma Grading

Zisheng Li$^{(\boxtimes)}$ and Masahiro Ogino

Research and Development Group, Hitachi, Ltd., Tokyo, Japan
zisheng.li.fj@hitachi.com

Abstract. In current oncological workflows of clinical decision making and treatment management, biopsy is the only way to confirm the abnormality of cancer. On the purpose of reducing unnecessary biopsies and diagnostic burden, we propose a patient-wise feature transfer model for learning the relationship of phenotypes between radiological images and pathological images. We hypothesize that high-level features from the same patient are possible to be linked between modalities of different image scales. We integrate multiple feature transfer blocks between CNN-based networks with single-/multi-modality radiological images and pathological images in an end-to-end training framework. We refer to our method as "augmented radiology" because the inference model only requires radiological images as input while the prediction result can be linked to specific pathological phenotypes. We apply the proposed method to glioma grading (high-grade vs. low-grade) and train the feature transfer model by using patient-wise multimodal MRI images and pathological images. Evaluation results show that the proposed method can achieve pathological tumor grading score in high accuracy (AUC 0.959) only given the radiological images as input.

Keywords: Augmented radiology · Feature transfer · Pathological prediction · Deep learning

1 Introduction

In current oncological workflows, radiology imaging techniques like computed tomography (CT), X-ray, and MRI can provide thorough view of a tumor and be used to monitor the progression of the tumor. However, biopsy and the corresponding pathological imaging is the only way to confirm the abnormality of tumor. Therefore, doctors have to consult cross-modality data such as radiological data, pathological data, and so on, when diagnosing a cancer disease and making treatment decisions. On the other hand, biopsy and pathological imaging brings diagnostic burden to doctors and patients and lowers the throughput of diagnosis workflow. Recently, radiomics methods are developed to help reducing diagnostic burden and clinical cost [1, 2]. These methods extract handcrafted image features which are related to genomic data, such as gene expression profiling and genotyping. However, such handcrafted features are not comprehensive, and the image features and the genomic features are difficult to be analyzed in a uniformed feature space.

S. Albarqouni et al. (Eds.): DART 2020/DCL 2020, LNCS 12444, pp. 23–30, 2020.
https://doi.org/10.1007/978-3-030-60548-3_3

In this paper, we propose a concept of "augmented radiology" based on a hypothesis that high-level features of radiological images and pathological images are possible to be linked and analyzed in the same feature space. We propose a feature transfer model for learning the relationship between modalities of different scales and resolutions. As a result, features of the radiological images are expected to be augmented by linking to that of the pathological images, and prediction only with the radiological images is expected to be improved. We apply the proposed method to glioma grade prediction (high-grade vs. low-grade) and train the feature transfer model by using patient-wise multimodal MRI images and pathological images.

Gliomas are the most frequent malignant primary brain tumors in adults, and it can be classified into four grade levels. Glioma patients of grade II and grade III are classified into low-grade status, and patients of grade IV are classified into high-grade status. Since it is very challenging for diagnosing different grade levels of glioma by using MRI screening, recently, there are a lot of novel works proposed to predict glioma grades by analyzing brain MRI images. In [3], radiomics features are manually extracted from multimodal MRI images, and a random forest algorithm is used to predict glioma grading. In [4], an adversarial-network-based method is proposed to synthesize multimodal MRI images and predict glioma grading by using single-modality MRI image input. In [5], a prediction model based on 3D DenseNet [6] is proposed to predict IDH genotypes and glioma grading. These methods only utilize MRI images as image data for training the prediction models. In [7], a concept of Mammography-Histology-Phenotype-Linking-Model is proposed for learning the relationship of features between mammography images and pathology images of breast cancer. However, detailed implementation or evaluation are not given, and more importantly, the proposed concept requires strict registration between image patches of mammography images and pathology images, which is difficult to been applied in practical workflow of cancer diagnosis and therapy.

Comparing with the previous literatures, contributions of our work include:

- We propose a feature transfer model that can learn the relationship between modalities of different scales and resolutions without image registration information.
- The proposed training framework is end-to-end, and the architecture of multiple feature transfer blocks can be easily be implemented to any CNN-based networks.
- The prediction model only requires radiological images as input for inference. And more accurate glioma grading can be obtained, compared to the prediction model without feature transfer learning.

2   Materials and Methodology

2.1   Datasets and Preprocessing

We use image data from the Cancer Genome Atlas (TCGA) to predict grading of brain glioma with two pathologic types: Lower Grade Glioma (LGG [8]) of Grade II and Grade III, and Glioblastoma Multiforme (GBM [9]) of Grade IV. For radiological images, we use brain MR images of T1 postcontrast, T1, T2 and Flair modalities. And we use tumor masks provided by BRAin Tumor Segmentation challenge (BRATS 2017) [10–12] to generate tumor image ROIs as dataset for training and test. All MR images and the corresponding masks are cropped and resampled to size of $112 \times 142 \times 112$ with resolution of $1.0 \times 1.0 \times 1.0 \ \text{mm}^3$. For each patient, we extract 32 tumor ROIs with size of 112×142 from axial slices around each tumor centroid.

We use whole-slide pathology images from the corresponding patients in LGG dataset and GBM dataset for the feature transfer learning. For each patient, we extract 32 image tiles with size of 256×256 at 20x magnification. On the purpose of extracting local representative image tiles from whole-slide digital pathology images, we proposed a preprocessing pipeline as shown in Fig. 1. First, we extract meaningful regions for the whole-slide pathology image by using color thresholding. As a result, regions of white background and stained pollution can be excluded. Then we randomly extract image tiles with size of 256×256 at 20x magnification from the remained regions. Next, we perform staining unmixing [13] by separating the hematoxylin and eosin stains on each image tile.

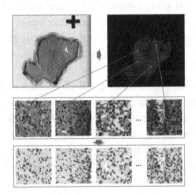

Fig. 1. Preprocessing pipeline of whole-slide digital pathology image.

2.2   Baseline Radiological Model

As a baseline method for tumor grading prediction by using radiological images only, we propose an end-to-end CNN-based model as shown in Fig. 2. We adopt VGG-16 models [14] to extract intermediate-level image features from each tumor-ROI input of multi-modal brain MRI images. The initial weights of the VGG-16 network are pre-trained by ImageNet data [15]. We then concatenate the intermediate-level features and extract high-level features by applying two fully-connected (FC) layers. Tumor grading is predicted by using the obtained high-level features from the multi-modal MRI images. The proposed prediction model is 2D-based, and the tumor grading is estimated by the ensemble prediction results of multiple (32) tumor ROIs from the same patient.

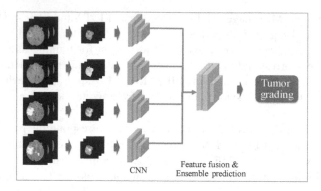

Fig. 2. Baseline radiological prediction model.

2.3   Proposed Method: Augmented Radiological Model

The proposed end-to-end training framework is shown in Fig. 3. The overall network consists of two prediction networks for radiological images and pathological images, respectively. The radiological image network has the same architecture as that of the baseline model. The pathological image network also has a similar architecture as that of the radiological image network. For an input of pathological image tile, a VGG-16 CNN network followed by two fully-connected dense layers are utilized to extract image features and perform tumor grading prediction.

We propose feature transfer (FT) blocks, which consist of layers of global pooling, normalization, and fully-connected, to transfer image features from pathological image domain to radiological image domain. We insert 4 FT blocks between the "block3_-pool" layers of every two VGG-16 networks from pathology and radiology domains. The transferred features of the radiology modalities are fed back to their own CNN networks. We also insert a similar FT block (but without global pooling) between the FC layers of the radiological image network and pathological image network.

At each iteration of the training process, radiological tumor ROIs and pathological image tiles from a same patient are randomly selected and input to the CNN networks. Ensemble predictions of radiological model and pathological model are performed by averaging multiple prediction results of different tumor ROIs and pathological image tiles, respectively. Registration information between the radiological images and pathological images are not necessary for the training because the relationship learning is performed on intermediate-level and high-level image feature space.

We use binary crossentropy and mse as loss function for prediction and feature transfer, respectively. The overall loss function is as follows:

$$\mathcal{L} = \lambda_{\text{Rad}} \cdot \mathcal{L}_{\text{Rad}} + \lambda_{\text{Path}} \cdot \mathcal{L}_{\text{Path}} + \lambda_{\text{FT}} \cdot \mathcal{L}_{\text{FT}} \tag{1}$$

where $\mathcal{L}_{\text{Rad}}$ and $\mathcal{L}_{\text{Path}}$ represent prediction losses of radiological image prediction and pathological image prediction, respectively. And λ_{Rad} and λ_{Path} are the corresponding loss weights. $\mathcal{L}_{\text{FT}}$ and λ_{FT} represent the loss function and loss weight of feature transfer. Considering scale difference of crossencropy and mse, we set the loss weights as

$\lambda_{\text{Rad}} = 0.4$, $\lambda_{\text{Path}} = 0.5$ and $\lambda_{\text{FT}} = 0.1$. The feature transfer loss $\mathcal{L}_{\text{FT}}$ consists of mse distance losses of intermediate-level features ($\mathcal{L}_{\text{T1Gd}}, \mathcal{L}_{\text{T1}}, \mathcal{L}_{\text{T2}}, \mathcal{L}_{\text{Flair}}$) and high-level features $\mathcal{L}_{\text{FC}}$:

$$\mathcal{L}_{\text{FT}} = \lambda_{\text{inter}} \cdot (\mathcal{L}_{\text{T1Gd}} + \mathcal{L}_{\text{T1}} + \mathcal{L}_{\text{T2}} + \mathcal{L}_{\text{Flair}}) + \lambda_{\text{FC}} \cdot \mathcal{L}_{\text{FC}} \tag{2}$$

where $\lambda_{\text{inter}} = 0.125$, $\lambda_{\text{FC}} = 0.5$ are set experimentally.

For inference, only the tumor ROIs of multi-modal MRI images are used as input, and pathology-like predicting result can be obtained.

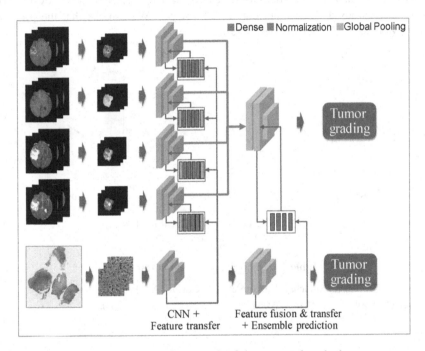

Fig. 3. Training framework of the proposed method.

3 Experiments and Results

3.1 Training/Evaluation Details

In the evaluation experiments, as radiological image input, we use multi-modal MRI images from the training set and the test set of BRATS 2017 as training data and test data, respectively. The corresponding patients' pathological images are extracted from TCGA. 80% of image data in the training set is used for training and 20% is for validation. As a result, we use 134 patients, 33 patients and 76 patients for processes of training, validation and test.

The proposed method is implemented in Tensorflow with Keras backend. All the experiments are run on a 24 GB NVIDIA Titan RTX GPU. In the training stages of all models, we use an Adam optimizer with a learning rate of 10^{-5}. Models that obtained the best validation loss are applied for test data.

We refer to the baseline model for radiological image as $Model_{Rad}$, the prediction models for radiological images and pathological images in the proposed feature transfer model as $Model_{FT}$ and $Model_{Path}$. ROC curves of the three models are shown in Fig. 4. We can see that can $Model_{Path}$ can achieve high accuracy (AUC 0.993) of low- and high- grade prediction of brain glioma, which guarantees the potential accuracy of $Model_{FT}$. Moreover, the feature transfer learning of $Model_{FT}$ is effective and improves the prediction accuracy of radiological feature from AUC 0.906 to AUC 0.959. For comparison, we also list the state-of-art results in Table 1. Note that CoCa-GAN's result is in accuracy and the input data is T1 images. We can see that the proposed method ($Model_{FT}$) achieves the best results.

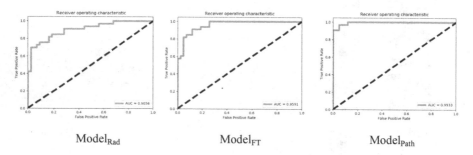

$Model_{Rad}$ $Model_{FT}$ $Model_{Path}$

Fig. 4. ROC curves of different models.

Table 1. Comparison prediction results.

	$Model_{Rad}$	$Model_{FT}$	Radiomics [3]	CoCa-GAN [4]	3D DenseNet [5]
AUC	0.906	**0.959**	0.921	0.879 (ACC)	0.948

Table 2. Results of $Model_{Rad}$ and $Model_{FT}$ on different number of training data.

Percent of training data	AUC of $Model_{Rad}$	AUC of $Model_{FT}$
100%	0.906	0.959
75%	0.903	0.958
50%	0.878	0.958
25%	0.781	0.911

On the purpose of evaluating training performance of the proposed feature transfer model, we reduce the number of training set and obtain AUC scores on the same test set. Results are shown in Table 2. We can see that when the data number of training set decreases to a small number (25% of the original training set), accuracy of baseline prediction model on radiological image drops from AUC 0.906 to AUC 0.781.

However, the proposed method can provide correct guidance for the training and obtain a good accuracy of AUC 0.911.

We also visualize the image features obtained by the baseline $Model_{Rad}$ and the proposed $Model_{FT}$ to confirm the effectiveness of the feature transfer learning. The results are shown in Fig. 5 and Fig. 6. In Fig. 5, heatmaps of intermediate-level features are plotted. We can see that regions that help for tumor grading are highlighted after the image feature is transferred in $Model_{FT}$. In Fig. 6, high-level features of the FC layers extracted from the test data are compressed with t-SNE. We can see that radiological image features after transfer learning distribute similar with the pathological ones. We consider that such feature transfer will help improving the prediction performance of radiological images to a large extent.

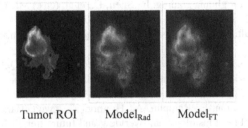

Tumor ROI $Model_{Rad}$ $Model_{FT}$

Fig. 5. Heatmaps of intermediate-level features.

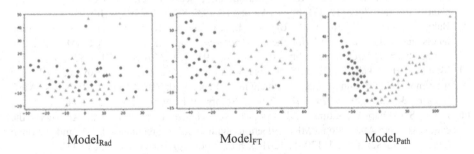

$Model_{Rad}$ $Model_{FT}$ $Model_{Path}$

Fig. 6. Visualization of high-level features by using t-SNE.

4   Conclusion

We propose a feature transfer model that can learn the relationship between modalities of different scales and resolutions without image registration information. The proposed training framework is end-to-end, and the architecture of multiple feature transfer blocks can be easily be implemented to any CNN-based networks. The transfer model can predict pathological tumor grading score in high accuracy (AUC 0.959) only given the radiological images as input. Moreover, pathological image features can appropriately guide the training of the prediction model even with a small (25%) training data number. The transfer model is expected to provide pathological predictive information for decision making of biopsy and treatment management by only using radiological images.

References

1. Beig, N., et al.: Radiogenomic analysis of hypoxia pathway is predictive of overall survival in Glioblastoma. Sci. Rep. **8**(1), 1–11 (2018)
2. Barbash, I.M., Waksman, R.: Current status, challenges and future directions of drug-eluting balloons. Future Cardiol. **7**(6), 765–774 (2011)
3. Cho, H.H., Lee, S.H., Kim, J., Park, H.: Classification of the glioma grading using radiomics analysis. PeerJ **6**, e5982 (2018). https://doi.org/10.7717/peerj.5982
4. Huang, P., et al.: CoCa-GAN: common-feature-learning-based context-aware generative adversarial network for glioma grading. In: Shen, D., et al. (eds.) MICCAI 2019. LNCS, vol. 11766, pp. 155–163. Springer, Cham (2019). https://doi.org/10.1007/978-3-030-32248-9_18
5. Liang, S., et al.: Multimodal 3D DenseNet for IDH genotype prediction in gliomas. Genes **9** (8), 382 (2018)
6. Huang, G., Liu, Z., Van Der Maaten, L., Weinberger, K.Q.: Densely connected convolutional networks. In: Proceedings of the IEEE Conference on Computer Vision and Pattern Recognition, pp. 4700–4708 (2017)
7. Hamidinekoo, A., Denton, E., Rampun, A., Honnor, K., Zwiggelaar, R.: Deep learning in mammography and breast histology, an overview and future trends. Med. Image Anal. **47**, 45–67 (2018)
8. Bakas, S., Akbari, H., Sotiras, A., Bilello, M., Rozycki, M., Kirby, J.S., et al.: Segmentation labels and radiomic features for the pre-operative scans of the TCGA-LGG collection. Cancer Imaging Arch. (2017). https://doi.org/10.7937/K9/TCIA.2017.GJQ7R0EF
9. Bakas, S., Akbari, H., Sotiras, A., Bilello, M., Rozycki, M., Kirby, J.S., et al.: Segmentation labels and radiomic features for the pre-operative scans of the TCGA-GBM collection. Cancer Imaging Arch. (2017). https://doi.org/10.7937/K9/TCIA.2017.KLXWJJ1Q
10. Menze, B.H., Jakab, A., Bauer, S., Kalpathy-Cramer, J., Farahani, K., Kirby, J., et al.: The multimodal brain tumor image segmentation benchmark (BRATS). IEEE Trans. Med. Imaging **34**(10), 1993–2024 (2015). https://doi.org/10.1109/TMI.2014.2377694
11. Bakas, S., Akbari, H., Sotiras, A., Bilello, M., Rozycki, M., Kirby, J.S., et al.: Advancing the cancer genome Atlas glioma MRI collections with expert segmentation labels and radiomic features. Nat. Sci. Data **4**, 170117 (2017). https://doi.org/10.1038/sdata.2017.117
12. Bakas, S., Reyes, M., Jakab, A., Bauer, S., Rempfler, M., Crimi, A., et al.: Identifying the best machine learning algorithms for brain tumor segmentation, progression assessment, and overall survival prediction in the BRATS challenge. arXiv preprint arXiv:1811.02629 (2018)
13. Macenko, M., et al.: A method for normalizing histology slides for quantitative analysis. In: 2009 IEEE International Symposium on Biomedical Imaging: From Nano to Macro, pp. 1107–1110 (2009)
14. Simonyan, K., Zisserman, A.: Very deep convolutional networks for large-scale image recognition. arXiv preprint arXiv:1409.1556 (2014)
15. Deng, J., Dong, W., Socher, R., Li, L.J., Li, K., Fei-Fei, L.: ImageNet: a large-scale hierarchical image database. In 2009 IEEE Conference on Computer Vision and Pattern Recognition, pp. 248–255 (2009)

Attention-Guided Deep Domain Adaptation for Brain Dementia Identification with Multi-site Neuroimaging Data

Hao Guan, Erkun Yang, Pew-Thian Yap, Dinggang Shen$^{(\boxtimes)}$, and Mingxia Liu$^{(\boxtimes)}$

Department of Radiology and BRIC, University of North Carolina at Chapel Hill, Chapel Hill, NC 27599, USA
Dinggang.Shen@gmail.com, mxliu@med.unc.edu

Abstract. Deep learning has demonstrated its superiority in automated identification of brain dementia based on neuroimaging data, such as structural MRIs. Previous methods typically assume that multi-site data are sampled from the same distribution. Such an assumption may not hold in practice due to the data heterogeneity caused by different scanning parameters and subject populations in multiple imaging sites. Even though several deep domain adaptation methods have been proposed to mitigate data heterogeneity between sites, they usually require a portion of labeled target data for model training, and rarely consider the potentially different contributions of different brain regions to disease prognosis. To address these limitations, we propose an attention-guided deep domain adaptation (AD^{2}A) framework for brain dementia prognosis, which does not need label information of the target domain and can automatically identify discriminative locations in whole-brain MR images. The proposed AD^{2}A framework consists of three key components: 1) a feature encoding module for representation learning of input MR images, 2) an attention discovery module for automatically locating dementia-related discriminative regions in brain MRIs, and 3) a domain transfer module with adversarial learning for knowledge transfer between the source and target domains. Extensive experiments have been conducted on three benchmark neuroimaging datasets, with results suggesting the effectiveness of our method in both tasks of brain dementia identification and disease progression prediction.

1 Introduction

Structural magnetic resonance imaging (MRI) data acquired from multiple imaging sites have been widely used in the automated prognosis of brain dementia, such as Alzheimer's disease (AD) and its early stage, *i.e.*, Mild Cognitive Impairment (MCI) [1–5]. Deep learning has recently demonstrated its superiority over conventional machine learning methods in neuroimaging-based prognosis of brain

© Springer Nature Switzerland AG 2020
S. Albarqouni et al. (Eds.): DART 2020/DCL 2020, LNCS 12444, pp. 31–40, 2020.
https://doi.org/10.1007/978-3-030-60548-3_4

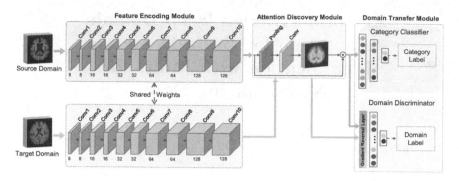

Fig. 1. Illustration of our attention-guided deep domain adaptation (AD^2A) framework for MRI-based disease identification.

dementia [5,6]. Previous studies typically assume that multi-site neuroimaging data are sampled from the same distribution [5–7], while such an assumption is too strong and may not hold in practice due to the data heterogeneity caused by different scanning parameters and subject populations in multiple sites.

To tackle the challenge of data heterogeneity in multi-domain neuroimaging data, several machine-learning-based domain adaptation methods have been proposed to minimize domain boundaries [8]. They often use shallow models that are trained using hand-crafted MRI features. To leverage more descriptive features, deep learning models, such as convolutional neural networks (CNNs), have been introduced to enhance the transferability of learned models between the source and target domains [9] and to handle inter-domain classification of brain dementia [10]. However, the existing methods often suffer from the following limitations. 1) Many of them need *a part of labeled target data* for model training, which limits their applications to unsupervised scenarios without any labeled target data. 2) They rarely explore the *potentially different contributions of different brain regions* in brain MR images to build robust prognostic models.

To this end, we propose an attention-guided deep domain adaptation (AD^2A) model for brain dementia prognosis. As shown in Fig. 1, the proposed AD^2A consists of three key components: 1) a *feature encoding module* that extracts feature representations of the input MR images in both source and target domains, 2) an *attention discovery module* that discovers disease-related regions in MRIs, and 3) a *domain transfer module* with adversarial learning for knowledge transfer between the source and target domains. The major contributions of this work is two-fold. *First*, our proposed model is unsupervised, without requiring any labeled data in the target domain. *Besides*, we explicitly incorporate discriminative region localization into the domain adaptation process, which will help reduce the negative influence of those uninformative brain regions for prognosis.

2 Methodology

2.1 Problem Definition

We study the problem of unsupervised domain adaptation for MRI-based brain dementia prediction. Let $\mathcal{X} \times \mathcal{Y}$ denote the joint space of samples and their corresponding class labels. A source domain S and a to-be-analyzed target domain T are defined on the joint space, with unknown and different distributions P and Q (P $\neq$ Q) respectively. Suppose we have n_s samples (subjects) with class labels in the source domain, i.e., $\mathcal{D}_S = \{(\mathbf{x}_i^S, y_i^S)\}_{i=1}^{n_s}$, and n_t samples in the target domain without class labels, i.e., $\mathcal{D}_T = \{(\mathbf{x}_j^T)\}_{j=1}^{n_t}$. We assume the two domains share the same set of class labels. Then the goal is to design a learning model that can precisely estimate the labels of subjects in the target domain, and such a model is constructed based on labeled samples in the source domain.

2.2 Proposed Attention-Guided Deep Domain Adaptation (AD²A)

Feature Encoding. We exploit a convolutional neural network as the backbone to extract features of brain MR images in the source and target domains. As shown in Fig. 1, the backbone network in the proposed AD²A contains ten $3\times3\times3$ convolution (Conv) layers. The numbers of channels for these sequential layers are 8, 8, 16, 16, 32, 32, 64, 64, 128, and 128, respectively. All of the Conv layers are followed by batch normalization operation and rectified linear unit (ReLU). To reduce over-fitting, down-sampling operations with the stride of $2 \times 2 \times 2$ are added in the Conv2, Conv4, Conv6, Conv8 and Conv10, respectively.

Dementia Attention Discovery. It has been revealed that dementia diseases are relevant with certain brain regions [11–13]. Accordingly, we design a trainable dementia attention module to automatically identify brain regions that are related to subject-specific dementia status in brain MR images. Feature map generated from the feature extraction network is fed into the attention block. Denote the input feature map as $\mathbf{M} = [\mathbf{M}_1, \cdots, \mathbf{M}_C]$, where $\mathbf{M}_i \in \mathbb{R}^{H \times W \times D} (i = 1, 2, ..., C)$ represents the feature map at the ith channel, and C denotes the number of channels. We then perform cross-channel average pooling and max pooling on $\mathbf{M}$ to generate two feature maps, i.e., $\mathbf{M}_{avg}$ and $\mathbf{M}_{max}$, respectively. These two feature maps are then concatenated, followed by a Conv layer (with one 3D Conv filter) to produce a spatial map. Finally we apply the sigmoid function as the nonlinear activation to compute the final attention map $\mathbf{A}$. Mathematically, the dementia attention map is calculated as:

$$\mathbf{A} = \sigma(f^{3\times3\times3}([\mathbf{M}_{max}, \mathbf{M}_{avg}])) \tag{1}$$

where σ is the sigmoid function and $f^{3\times3\times3}$ represents a convolution operator with a $3 \times 3 \times 3$ kernel. The sigmoid function is used to enforce each point in the generated attention map is between 0 and 1 which can play a role of weighting, i.e., disease-related areas will have higher weights than other areas. Our attention module uses both category labels and domain labels as supervision for training (see Fig. 1). This is different from previous deep learning-based ones [14] that are trained with only category labels as supervision. Therefore, our attention module can highlight discriminative regions across different domains, while others can only focus on a single domain.

Domain Transfer via Adversarial Learning. For unsupervised domain adaptation, due to the distribution difference between the source and target data, a model well-trained on a source domain may have poor robustness when applied to the target domain. Our goal is to construct a robust model (based on labeled source data) for the target domain. To this end, we introduce a domain transfer module based on adversarial learning to make a trade-off between classification performance and the generalization ability of the learning model. Specifically, the proposed module contains a category classifier and a domain discriminator/classifier, which are co-trained during the process of model learning. The category classifier is developed for disease identification, while the domain discriminator tells whether an input sample is from the source/target domain. Through co-training, these two classifiers encourage the model to *not only* achieve good classification performance for source data *but also* learn domain-invariant MRI features for both domains, thus improving the robustness of the learned model when applied to the target domain.

Category Classifier: The category classifier $\mathcal{C}_S$ is used to predict the labels of input samples. Since no label information is available for the target data, only labeled source data are used for training this classifier. With the features generated by the feature extractor network as input, two fully-connected layer with 32 and 2 units are used in the category classifier. The cross-entropy loss is used in $\mathcal{C}_S$ as:

$$\mathcal{L}(\mathcal{C}_S, \mathbf{X}^S, \mathbf{y}^S) = \frac{1}{n_s} \sum_{i=1}^{n_s} L(\mathcal{C}_S(\mathbf{x}_i^S), y_i^S), \tag{2}$$

where $L(\cdot)$ denotes the cross-entropy loss.

Domain Discriminator: The domain discriminator $\mathcal{C}_D$ is used to distinguish features from different domains. It actually serves as a player (another one is the feature encoding module) in a minmax game, *i.e.*, adversarial learning. Through trying to maximize the loss of the domain classifier, the feature encoding module is encouraged to learn domain-invariant MRI features for both source and target data. We add two fully-connected layers with 32 and 2 units in the domain discriminator. For input source and target samples, we form a training batch $\{(x_1, y_1^D), (x_2, y_2^D), \cdots, (x_N, y_N^D)\}$ with N samples, where $y_i^D = 1$ indicates that x_i comes from the source domain and $y_i^D = 0$ denotes that x_i is from the target domain. Equal numbers of training samples from both source domain and target domain are selected without bias towards either of them. The domain discriminator is learned by minimizing the cross-entropy loss as:

$$\mathcal{L}(\mathcal{C}_D, \mathbf{X}^D, \mathbf{y}^D) = \frac{1}{N} \sum_{j=1}^{N} L(\mathcal{C}_D(\mathbf{x}_j), y_j^D), \tag{3}$$

where $L(\cdot)$ denotes the cross-entropy loss and y_i^D is the domain label.

Since the final goal is to learn domain-invariant features across the source and target domain, the task can be performed by learning a model that is able to predict labels correctly without any domain cues. This can be achieved by jointly minimizing the category classification loss in Eq. 2 and maximizing the domain classification loss in Eq. 3 as:

$$\mathcal{L}_{total} = \mathcal{L}(\mathcal{C}_S, \mathbf{X}^S, \mathbf{y}^S) - \alpha \mathcal{L}(\mathcal{C}_D, \mathbf{X}^D, \mathbf{y}^D), \tag{4}$$

where α is used to control the contribution of the domain classification loss. This optimization of Eq. 4 can be solved by standard stochastic gradient descent (SGD). Since the parameters of the feature extraction network are jointly determined by the back-propagation of category classifier and domain discriminator, following [15], we add a self-defined *gradient reversal layer* to transmit negative gradient variations by the domain discriminator, as shown in Fig. 1.

Implementation. The proposed model was implemented in PyTorch. The network was trained for 100 epochs. The Adam optimizer was used with a learning rate of 1×10^{-4} and a batch size of 2. To prevent over-fitting, a dropout operation with a rate of 0.5 was used. The parameter α in Eq. 4 was empirically set to 0.01. Training was divided to two stages. Firstly, the feature encoding and attention discovery modules were pre-trained with the category classifier via Eq. 2. Then they were fine-tuned and co-trained with both the domain discriminator and category classifier via Eq. 4.

3   Experiments

Materials and MR Image Pre-processing. Three benchmark datasets are used for evaluation, *i.e.*, Alzheimer's Disease Neuroimaging Initiative (ADNI-1) [16], ADNI-2, and Australian Imaging Biomarkers and Lifestyle Sudy of Aging database (AIBL) [17]. Subjects that appear in both ADNI-1 and ADNI-2 are removed from ADNI-2 for independent evaluation. ADNI-1 contains 1.5T T1-weighted structural MRIs acquired from 748 subjects, including 205 AD, 231 cognitively normal (CN), 165 progressive MCI (pMCI) and 147 stable MCI (sMCI) subjects. ADNI-2 consists of 3.0T T1-weighted structural MRIs acquired from 708 subjects, including 162 AD, 205 CN, 88 pMCI and 253 sMCI subjects. AIBL contains MRIs acquired from 549 subjects, including 71 AD, 447 CN, 11 pMCI and 20 sMCI subjects. All brain MRIs were pre-processed through a standard pipeline, including skull stripping, intensity correction, and spatial normalization. Finally, all MRIs were cropped and normalized to have the identical size of $181 \times 217 \times 181$, under the requirement that all brain tissues should be completely preserved without loss of any useful information.

Experimental Setup. Two groups of experiments were performed, including: 1) AD identification (*i.e.*, AD vs. CN classification), and 2) MCI conversion

Table 1. Results of five methods in the task of brain dementia identification (*i.e.*, AD vs. CN classification) in four different transfer learning settings.

Task (Source → Target)	Method	ACC (%)	SEN (%)	SPE (%)	AUC (%)
ADNI-1 → ADNI-2	TCA	74.14	54.72	88.55	80.57
	SA	74.44	62.26	84.08	80.21
	GFK	63.06	53.46	70.65	69.18
	DANN	87.14	75.68	90.61	90.81
	AD^{2}A (Ours)	**89.10**	**81.48**	**95.12**	**93.35**
ADNI-2 → ADNI-1	TCA	73.60	66.83	79.48	80.59
	SA	73.36	59.30	85.59	80.17
	GFK	62.15	49.25	73.36	65.03
	DANN	84.03	76.19	85.73	89.24
	AD^{2}A (Ours)	**87.39**	**86.83**	**87.88**	**91.01**
ADNI-1 → AIBL	TCA	68.38	31.65	77.20	50.95
	SA	69.61	36.58	79.94	51.34
	GFK	59.80	46.27	35.44	56.53
	DANN	86.55	73.24	90.00	90.15
	AD^{2}A (Ours)	**88.30**	**84.51**	**91.00**	**92.50**
ADNI-1+ADNI-2 → AIBL	TCA	69.36	28.99	77.51	51.68
	SA	74.02	26.46	85.84	52.33
	GFK	63.73	34.18	70.82	50.49
	DANN	88.89	80.32	92.00	93.12
	AD^{2}A (Ours)	**90.06**	**81.69**	**96.00**	**95.16**

prediction (*i.e.*, pMCI vs. sMCI classification). For AD identification, we considered four transfer learning settings: 1) "ANDI-1→ADNI-2" (with ADNI-1 as the source domain and ADNI-2 as the target domain), 2) "ANDI-2→ADNI-1", 3) "ANDI-1→AIBL", and 4) "ADNI-1+ADNI-2 → AIBL" (with the combination of ADNI-1 and ADNI-2 as the source domain and AIBL as the target domain). Since AIBL has too few MCI subjects which are not sufficient to train a good network, for MCI conversion prediction, we only considered two transfer learning settings: 1) "ANDI-1→ADNI-2", and 2) "ANDI-2→ADNI-1". Four metrics were used, *i.e.*, classification accuracy (ACC), sensitivity (SEN), specificity (SPE), and area under receiver operating characteristic curve (AUC).

We compared our AD^{2}A with 3 conventional methods using hand-crafted features (*i.e.*, gray matter volumes of 90 regions defined in the AAL template), including 1) Transfer Component Analysis (TCA) [18], 2) Subspace Alignment (SA) [19], and 3) Geodesic Flow Kernel [20], as well as 4) the state-of-the-art Domain-Adversarial Training of Neural Network (**DANN**) [15]. For the three conventional methods, logistic regression was used as the classifier. Default parameters in their respective papers are used in the competing methods.

Table 2. Results of five methods in the task of MCI conversion prediction (*i.e.*, pMCI vs. sMCI classification) in two different transfer learning settings.

Task (Source → Target)	Method	ACC (%)	SEN (%)	SPE (%)	AUC (%)
ADNI-1 → ADNI-2	TCA	70.76	36.84	76.15	62.85
	SA	66.06	28.95	71.97	57.81
	GFK	69.68	34.21	75.31	55.75
	DANN	74.78	51.55	81.22	75.32
	AD^{2}A (Ours)	**77.42**	**52.27**	**86.17**	**78.20**
ADNI-2 → ADNI-1	TCA	58.27	53.89	61.50	60.01
	SA	58.78	58.68	58.85	59.88
	GFK	53.94	44.31	61.06	50.88
	DANN	67.31	58.18	73.55	66.32
	AD^{2}A (Ours)	**68.91**	**64.24**	**74.15**	**69.51**

Evaluation of Disease Identification. Results achieved by different methods in the task of dementia identification are reported in Table 1. From Table 1, we have the following two observations. *First*, the proposed AD^{2}A achieves consistently better performance than the conventional and deep learning methods in four transfer learning settings. *Second*, our AD^{2}A yields overall better results in the setting of "ADNI-1+ADNI-2 → AIBL" than "ADNI-1 → AIBL". This indicates that using more diverse data in the source domain may improve the robustness of the learned models when applied to the unlabeled target domain. In addition, we have performed an experiment on the merged ADNI-1 and ANDI-2 data (with 20% for test, 10% for validation, and 70% for training), and recorded AUCs of 5 methods in AD vs. CN classification (Ours: 94.33%, TCA: 82.24%, SA: 83.09%; GFK: 78.60%; DANN: 91.82%).

Evaluation of Disease Progression Prediction. Results achieved by different methods in MCI convert prediction are reported in Table 2. As can be seen from Table 2, the performance of five methods in "ADNI-2 → ADNI-1" is usually worse than "ADNI-1 → ADNI-2". This should be caused by data imbalance in pMCI and sMCI subjects in ADNI-2. Besides, results in Tables 1–2 indicate that it is more challenging to accurately predict the conversion of MCI subjects than AD identification. Nevertheless, our AD^{2}A still outperforms the other methods.

Ablation Study. To evaluate the effectiveness of two major components (*i.e.*, attention discovery and domain discriminator), we compare AD^{2}A with its three variants: 1) ADN with only the feature encoding (FE) module and the category classifier; 2) ADN-T with the FE module, attention discovery module, and the category classifier; and 3) AD^{2}A S with the FE and domain transfer modules. Note that ADN and ADN-T first train models on source data and then directly apply the models to target data (*i.e.*, without domain adaptation). The ACC results achieved by our AD²A and its three variants are shown in Fig. 2, from which we can see that the attention component can improve the transferability of the classification system. *Second*, the domain adaptation module can further improve the performance for transfer classification. This can be explained by that it is able to learn domain-invariant features that reflect import locations of the brain for disease identification.

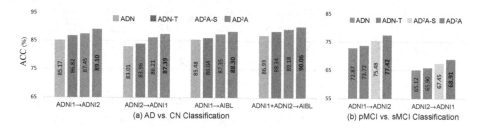

Fig. 2. Ablation study for verifying the effectiveness of different components in AD²A.

Contribution of Domain Discriminator. To study the contribution of the proposed domain discriminator in AD²A, we vary the value of α in Eq. 4 within the range of $[0, 0.01, 0.02, 0.05, 0.1, 0.2, 0.5, 1]$, and report the AUC values of AD²A. The experimental results of AD²A with different values of α in MCI conversion prediction in the setting of "ADNI-1 $\rightarrow$ ADNI-2" are reported in Fig. 3, from which one can observe that AD²A yields good results when $\alpha \in [0.01, 0.05]$. This suggests that the domain discriminator has a useful supplementary effect on AD²A in MCI conversion prediction.

Learned Attention Maps. We further visualize the attention maps generated by our AD²A for four typical subjects in Fig. 4. It can be seen that the most discriminative regions (denoted by red) mainly locate in the hippocampus [12] and ventricles [13], as well as the midbrain area (e.g., ventral tegmental area) in the brain stem, which is consistent with the more recent research [21,22]. This demonstrates that reserving the brain stem in MRI pre-processing is beneficial for intelligence identification of brain diseases.

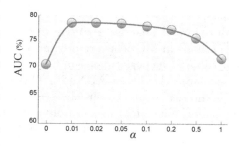

Fig. 3. Impact of the parameter α

Besides, we can observe that the discriminative regions of AD are more distinct on average than the regions of MCI (*i.e.*, pMCI and sMCI). Considering that the structural changes caused by AD are easier to detect than MCI, these results suggests that the attention maps learned by our AD²A are reasonable. An

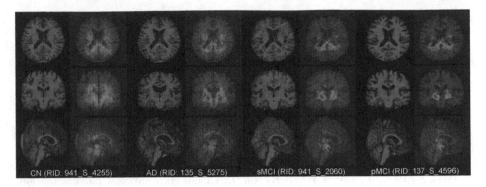

Fig. 4. Attention maps generated by our AD^2A for 4 typical subjects from ADNI-2. The red and blue denote the high and low discriminative capability of brain regions in disease identification, respectively. (Color figure online)

expert in our group with experience in diagnosis and treatment of Alzheimer's disease confirmed the rationality of attention maps generated by our model.

4 Conclusion

In this paper, we have proposed an attention-guided deep domain adaptation model to tackle the problem of inter-domain heterogeneity for brain dementia identification. Experimental results have offered the community some useful findings. The most important one is that attention-guided domain adaptation can more effectively learn domain-sharing and disease-associated features in brain MR images, thus greatly improve the transferability of machine learning models for neuroimaging analysis.

References

1. Falahati, F., Westman, E., Simmons, A.: Multivariate data analysis and machine learning in Alzheimer's disease with a focus on structural magnetic resonance imaging. J. Alzheimer's Dis. **41**(3), 685–708 (2014)
2. Cuingnet, R., Gerardin, E., Tessieras, J., et al.: Automatic classification of patients with Alzheimer's disease from structural MRI: a comparison of ten methods using the ADNI database. NeuroImage **56**(2), 766–781 (2011)
3. Liu, M., Zhang, D., Shen, D.: Relationship induced multi-template learning for diagnosis of Alzheimer's disease and mild cognitive impairment. IEEE Trans. Med. Imaging **35**(6), 1463–1474 (2016)
4. Lian, C., et al.: Multi-channel multi-scale fully convolutional network for 3D perivascular spaces segmentation in 7T MR images. Med. Image Anal. **46**, 106–117 (2018)
5. Lian, C., Liu, M., Zhang, J., Shen, D.: Hierarchical fully convolutional network for joint atrophy localization and Alzheimer's disease diagnosis using structural MRI. IEEE Trans. Pattern Anal. Mach. Intell. **42**(4), 880–893 (2018)

6. Liu, M., Zhang, J., Adeli, E., Shen, D.: Landmark-based deep multi-instance learning for brain disease diagnosis. Med. Image Anal. **43**, 157–168 (2018)
7. Valiant, L.G.: A theory of the learnable. Commun. ACM **27**(11), 1134–1142 (1984)
8. Wachinger, C., Reuter, M.: Domain adaptation for Alzheimer's disease diagnostics. NeuroImage **139**, 470–479 (2016)
9. Rieke, J., Eitel, F., Weygandt, M., Haynes, J.-D., Ritter, K.: Visualizing convolutional networks for MRI-based diagnosis of Alzheimer's disease. In: Stoyanov, D., et al. (eds.) MLCN/DLF/IMIMIC -2018. LNCS, vol. 11038, pp. 24–31. Springer, Cham (2018). https://doi.org/10.1007/978-3-030-02628-8_3
10. Hosseini-Asl, E., Keynton, R., El-Baz, A.: Alzheimer's disease diagnostics by adaptation of 3D convolutional network. In: ICIP, pp. 126–130 (2016)
11. Woo, S., Park, J., Lee, J.Y., So Kweon, I.: Cbam: convolutional block attention module. In: ECCV, pp. 3–19 (2018)
12. Mu, Y., Gage, F.H.: Adult hippocampal neurogenesis and its role in Alzheimer's disease. Mol. Neurodegeneration **6**(1), 85 (2011). https://doi.org/10.1186/1750-1326-6-85
13. Ott, B.R., Cohen, R.A., Gongvatana, A., et al.: Brain ventricular volume and cerebrospinal fluid biomarkers of Alzheimer's disease. J. Alzheimers Dis. **20**(2), 647–657 (2010)
14. Zhou, B., Khosla, A., Lapedriza, A., Oliva, A., Torralba, A.: Learning deep features for discriminative localization. In: CVPR, pp. 2921–2929 (2016)
15. Ganin, Y., Ustinova, E., Ajakan, H., et al.: Domain-adversarial training of neural networks. J. Mach. Learn. Res. **17**(1), 2030–2096 (2016)
16. Jack Jr., C.R., Bernstein, M.A., Fox, N.C., et al.: The Alzheimer's disease neuroimaging initiative (ADNI): MRI methods. J. Magn. Reson. Imaging **27**(4), 685–691 (2008)
17. Ellis, K.A., Bush, A.I., Darby, D., et al.: The Australian Imaging, Biomarkers and Lifestyle (AIBL) study of aging: methodology and baseline characteristics of 1112 individuals recruited for a longitudinal study of Alzheimer's disease. Int. Psychogeriatr. **21**(4), 672–687 (2009)
18. Pan, S.J., Tsang, I.W., Kwok, J.T., Yang, Q.: Domain adaptation via transfer component analysis. IEEE Trans. Neural Networks **22**(2), 199–210 (2010)
19. Fernando, B., Habrard, A., Sebban, M., Tuytelaars, T.: Unsupervised visual domain adaptation using subspace alignment. In: ICCV, pp. 2960–2967 (2013)
20. Gong, B., Shi, Y., Sha, F., Grauman, K.: Geodesic flow kernel for unsupervised domain adaptation. In: CVPR, pp. 2066–2073 (2012)
21. D'Amelio, M., Puglisi-Allegra, S., Mercuri, N.: The role of dopaminergic midbrain in Alzheimer's disease: translating basic science into clinical practice. Pharmacol. Res. **130**, 414–419 (2018)
22. De Marco, M., Venneri, A.: Volume and connectivity of the ventral tegmental area are linked to neurocognitive signatures of Alzheimer's disease in humans. J. Alzheimers Dis. **63**(1), 167–180 (2018)

Registration of Histopathology Images Using Self Supervised Fine Grained Feature Maps

James Tong[1,2(✉)], Dwarikanath Mahapatra[3(✉)], Paul Bonnington[2],
Tom Drummond[2], and Zongyuan Ge[1,2(✉)]

[1] Monash University, Melbourne, Australia
jmton7@student.monash.edu, zongyuan.ge@monash.edu
[2] Airdoc Research, Melbourne, Australia
[3] Inception Institute of Artificial Intelligence, Abu Dhabi, UAE
dwarikanath.mahapatra@inceptioniai.org

Abstract. Image registration is an important part of many clinical workflows and inclusion of segmentation information of structures of interest improves registration performance. We propose to integrate segmentation information in a registration framework using fine grained feature maps obtained in a self supervised manner. Self supervised feature maps enables use of segmentation information despite the unavailability of manual segmentations. Experimental results show our approach effectively replaces manual segmentation maps and demonstrate the possibility of obtaining state of the art registration performance in real world cases where manual segmentation maps are unavailable.

Keywords: Fine grained segmentation · Registration · Histopathology

1 Introduction

An important part of a digital pathology image analysis workflow is the visual comparison of successive tissue sections enabling pathologists to evaluate histology and expression of multiple markers in a single area [11]. It has the potential to improve diagnostic accuracy and requires aligning all images to a common frame. Due to tissue processing and other procedures, certain sections may undergo elastic deformations leading to shape changes across slices. Thus it is essential to have a reliable registration algorithm that can address challenges such as: (i) differences in tissue appearance (a consequence of different dyes and sample preparation); (ii) high computing time due to large whole slide images (WSI); (iii) correcting complex elastic deformations, and (iv) ambiguous global unique features due to repetitive texture [1].

J. Tong and D. Mahapatra—Equal Contributions.

© Springer Nature Switzerland AG 2020
S. Albarqouni et al. (Eds.): DART 2020/DCL 2020, LNCS 12444, pp. 41–51, 2020.
https://doi.org/10.1007/978-3-030-60548-3_5

A comprehensive review on conventional image registration methods can be found in [21]. Although widely used, these methods are time consuming due to: 1) iterative optimization techniques; and 2) extensive parameter tuning. Deep learning (DL) methods can potentially overcome these limitations by using trained models to output registered images and deformation fields in much less time. In recent DL based image registration works, Sokooti et al. [27] propose RegNet using CNNs trained on simulated deformations to register a pair of unimodal images. Vos et al. [29] propose the deformable image registration network (DIR-Net) which outputs a transformed image non-iteratively without having to train on known registration transformations. Rohe et al. [25] propose SVF-Net that uses deformations obtained by registering previously segmented regions of interest (ROIs). The above methods are limited by the need of spatially corresponding patches [27,29] or being too dependent on training data. Balakrishnan et al. [4] learn a parameterized registration function from a collection of training volumes, and in [3] they improve on their method by adding segmentation information.

Hu et al. in [8] propose a paradigm where registration is posed as a problem of segmenting corresponding regions of interest (ROIs) across images. Lee et al. in [13] register anatomical structures by mapping input images to spatial transformation parameters using Image-and-Spatial Transformer Networks (ISTNs). Liu et al. [14] propose feature-level probabilistic model to regularize CNN hidden layers for 3D brain image alignment. Hu et al. [7] use a two-stream 3D encoder-decoder network and pyramid registration module for registration. In previous work we have proposed generative adversarial network (GAN) based approaches for registration [15–17].

It is a well established fact that registration and segmentation are interrelated and mutually complementary. For example, labeled atlas images are used via image registration for segmentation while segmentation maps provide extra information to aid in image registration and result evaluation. Methods combining registration and segmentation have been proposed using active-contours [31] and Bayesian methods [24] or Markov random field formulations [19,20]. Recent deep learning based approaches have used GANs [18] and a Deep Atlas [30] for joint registration and segmentation.

1.1   Contributions

While these methods show the advantages of integrating segmentation information with registration, they are dependent upon the availability of annotated segmentation maps during training. We propose a method to integrate structural information from self-supervised segmentation maps for registering histopathological images. Our approach does not require manual segmentation maps during training and test times. We propose a novel pretext task designed to recover simulated deformations, similar to the downstream task of image registration with segmentation information, which proves to be accurate and robust. The network trained to solve the pretext task is used to generate coarse feature maps of the new image pair to be registered, and clustering is used to obtain fine grained feature maps. This structural information is included in the registration framework by means of a loss function that captures the structural similarity between the image pair being registered.

2   Method

Figure 1 shows the workflow of our proposed approach. We first train the self supervised registration network (SSR-Net). The reference image (I_R) and floating image (I_F) are passed through SSR-Net to generate feature maps I_R^M and I_F^M. The maps and original images are input to a UNet style network that outputs a registration field used by a spatial transformer network (STN) to output the transformed image.

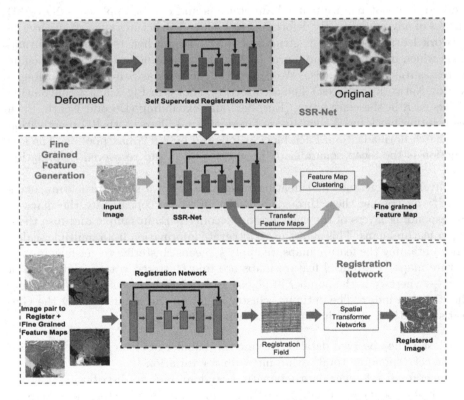

Fig. 1. Depiction of proposed workflow for training self supervised registration, generating fine grained feature maps and registration output.

2.1   Self Supervised Segmentation Feature Maps

Self-supervised learning is a variant of unsupervised learning methods, where a model learns to predict a set of labels that can be automatically created from the input data. A general approach is to train a CNN to perform an auxiliary task such as visual localization [12], predicting missing image parts [23], and cardiac MR image segmentation [2]. We train a UNet based registration network designed to predict the applied deformation field through simulated deformations on the image. Figure 1 (top row) shows an example of simulated deformation. First, by using watershed segmentation technique we identify different structures of interest (e.g., cells) in the histopathology image. Thereafter we select multiple structures of interest, deform them using freeform deformations [26] and change the intensity slightly by contrast stretching. We choose to deform well defined structures instead of random regions in the image since it aligns well with our main task of integrating segmentation with registration.

Since we know the applied deformation field, the loss function is the mean square error between the applied and recovered deformation fields. Our training method does not require manual annotations and can be applied to 2D or 3D images of any modality. By deforming well defined structures the self supervised network learns the inherent structural information when recovering deformations which improves performance. Intensity changes due to contrast stretching increases robustness to noise. We denote this network as Self-Supervised Registration Network (SSR-Net) The architecture, shown in Fig. 1 has 3 convolution blocks each in the contracting (Encoder) and expanding (Decoder) path. Each convolution block has 3 convolution layers of 3×3 filters, with ReLu activation and batch normalization. Each block is followed by 2×2 max pooling. The loss function is the mean square error between applied and recovered deformation field.

Given the reference and floating images I_R, I_F, we generate feature maps I_R^M, I_F^M by passing them through SSR-Net. First we concatenate the maps of corresponding layers of the encoder and decoder part, upsample and fuse them to get feature maps. This allows us to integrate information from multiple scales. After obtaining the feature maps we apply k-means clustering to get fine grained feature maps. The initial feature maps are first clustered with a high value of $k = \frac{3N}{4}$ where N is the number of pixels, and reduce it as $k_{t+1} = 0.9k_t$, t being the iteration index. The optimum cluster number is determined using the Gap statistic method of [28] having the following steps:

1. Cluster the observed data by varying the number of clusters k and compute the corresponding total within intra-cluster variation W_k.
2. Generate B reference data sets with a random uniform distribution. Cluster each of these reference data sets with varying number of clusters, and compute the corresponding total within intra-cluster variation W_{kb}.
3. Compute the estimated gap statistic as the deviation of the observed W_k value from its expected value W_{kb} under the null hypothesis: $Gap(k) = \frac{1}{B} \sum_{b=1}^{B} \log(W_{kb}^*) - \log(W_k)$. Also compute the standard deviation.
4. Choose the number of clusters as the smallest value of k such that the gap statistic is within one standard deviation of the gap at k+1, i.e., $Gap(k) \geq Gap(k+1) - s_{k+1}$.

2.2   Registration Using Segmentation Maps

Our registration approach is framework agnostic and can be used with various architectures. However, we choose to demonstrate our method's effectiveness using the VoxelMorph architecture [3] because it is a popular baseline registration method using segmentation information. They use a dice loss between manual segmentation maps in their training loss function. We investigate whether our self supervised fine grained feature maps can effectively replace the manual maps.

Figure 1 third row shows the registration network architecture. Given reference and floating images I_R, I_F, we assume they have been affinely aligned, and only non-linear displacements exist between them. A Encoder-Decoder architecture is used for computing the deformation field φ which is used to transform I_F and obtain the registered image $I_{Reg} = \varphi \circ I_F$ through spatial transformer networks (STNs) [9]. In an ideal registration I_R and I_{Reg} should match. Similar to the UNet architecture each block represents set of 3 2D convolution layers of kernel size 3, stride 2, followed by a LeakyReLU layer with parameter 0.2. The optimal parameter values are learned by minimizing differences between I_{Reg} and I_R . In order to use standard gradient-based methods, we construct a differentiable operation based on STN [9].

Loss Functions: We use two loss functions: an unsupervised loss L_{us} and an auxiliary loss L_a that leverages anatomical segmentations at training time. The unsupervised loss consists of two components: L_{sim} that penalizes differences in appearance, and L_{smooth} that penalizes local spatial variations in φ:

$$L_{us}(I_R, I_F, \varphi) = L_{sim}(I_R, I_{Reg}) + \lambda_1 L_{smooth}(\varphi) + \lambda_2 L_{seg}(I_R^M, I_F^M, I_{Reg}^M), \quad (1)$$

where $\lambda_1 = 0.95, \lambda_2 = 1.05$ are regularization parameters. L_{sim} is a combination of mean squared error (MSE) of pixel intensity difference, and local cross correlation (CC) between I_R, I_{Reg}. L_{seg} is the segmentation loss function and is defined as

$$L_{seg}(I_R^M, I_{Reg}^M) = MSE(I_R^M, I_{Reg}^M) \quad (2)$$

which is the pixel wise difference between the fine grained feature maps of the reference and registered image obtained from SSR-Net. $I_{Reg}^M = \varphi \circ I_F^M$ is obtained by applying the deformation field φ to the segmentation (or fine grained feature) map of the floating image being registered. A smooth displacement field φ is obtained using a diffusion regularizer on the spatial gradients of displacement u:

$$L_{smooth}(\varphi) = \sum_{p \in \Omega} \|\nabla u(p)\|^2 \quad (3)$$

3   Experimental Results

3.1   Implementation and Dataset Details

Our method was implemented in TensorFlow. We use Adam [10] with $\beta_1 = 0.93$ and batch normalization. The network was trained with 10^5 update iterations at learning rate 10^{-3}. Training and test was performed on a NVIDIA Titan X GPU with 12 GB RAM. We test our method on two datasets: 1) the publicly available ANHIR challenge dataset [1] used to evaluate automatic nonlinear image registration of 2D microscopy images of histopathology tissue samples stained with different dyes; and 2) a subset of the brain images used in [3]. Their descriptions are given below.

ANHIR Dataset: The dataset consists of 481 image pairs and is split into 230 training and 251 evaluation pairs [1]. For the training image pairs, both source and target landmarks are available, while for evaluation image pairs only source landmarks are available. The landmarks were annotated by 9 experts and there are on an average 86 landmarks per image. The average error between the same landmarks chosen by two annotators is 0.05% of the image size which can be used as the indicator of the best possible results to achieve by the registration methods, below which the results become indistinguishable [5].

The images are divided into 8 classes containing: (i) lesion tissue, (ii) lung lobes, (iii) mammary glands, (iv) the colon adenocarcinomas, (v) mice kidney tissues, (vi) gastric mucosa and adenocarcinomas tissues, (vii) breast tissues, (viii) human kidney tissues. All the tissues were acquired in different acquisition settings making the dataset more diverse. The challenge was performed on medium size images which were approximately 25% of the original image scale. The approximate image size after the resampling varied from $8k$ to $16k$ pixels (in one dimension).

Brain Image Dataset: We used the 800 images of the ADNI-1 dataset [22] consisting of 200 controls, 400 MCI and 200 Alzheimer's Disease patients. The MRI protocol for ADNI1 focused on consistent longitudinal structural imaging on $1.5T$ scanners using $T1$ and dual echo $T2-$weighted sequences. All scans were resampled to $256 \times 256 \times 256$ with 1mm isotropic voxels. Pre-processing includes affine registration and brain extraction using FreeSurfer [6], and cropping the resulting images to $160 \times 192 \times 224$. The dataset is split into 560, 120, and 120 volumes for training, validation, and testing. Registration is performed in 3D.

3.2   ANHIR Registration Results

Registration performance is evaluated using a normalized version of target registration error (TRE) and is defined as:

$$rTRE = \frac{TRE}{\sqrt{w^2 + h^2}} \tag{4}$$

where TRE is the target registration calculated as the Euclidean distance between corresponding landmarks in the two images and w, h are the image width and height respectively.

Table 1 summarizes the performance of $SR-Net$ (our proposed Segmentation based Registration NETwork) and the top 2 methods on different tissue types. The median rTRE for all tissues and each individual tissue type is reported with the numbers taken from [1].

We perform a set of ablation experiments where we exclude the segmentation loss of Eq. 2 (denoted as $SR - Net_{wL_{seg}}$), using either MSE or cross correlation in L_{sim} of Eq. 1 (denoted, respectively, $SR - Net_{MSE}$, $SR - Net_{MSE}$). Since our registration framework is similar to VoxelMorph, $SR - Net_{wL_{seg}}$ is equivalent to VoxelMorph without the segmentation loss. The results show that SR-Net outperforms the top ranked methods for the challenge. Moreover, the advantages of using segmentation is also obvious by the fact that the rTRE values for SR-$Net_{wL_{Seg}}$ are significantly higher than SR-Net ($p = 0.003$ from a paired Wilcoxon test). Ablation experiments also quantify the contribution of MSE and CC in the registration framework.

Figure 2 shows the registration results for pathology images where we show the reference and floating images alongwith the misalignment images before registration and after registration using SR-Net and SR-$Net_{wL_{Seg}}$. The misalignment is greatly reduced after registration using SR-Net while in the case of SR-$Net_{wL_{Seg}}$ there is still some resulting misalignment. This error can have significant consequences in the final diagnosis workflow. Hence the advantages of self-supervised segmentation maps are quite clear.

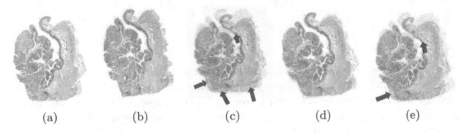

(a)   (b)   (c)   (d)   (e)

Fig. 2. Pathology image registration result: (a) I_R; (b) I_F; Misalignment of images: (c) before registration; after registration using (d) SR-Net; (e) SR-$Net_{wL_{Seg}}$

Table 1. Registration errors for different methods on the ANHIR dataset including ablation experiments. Values for *Rank 1, Rank 2* are taken from [1]

	$SR - Net$	Rank 1	Rank 2	SR-Net$_{wSeg}$	SR-Net$_{MSE}$	SR-Net$_{CC}$
All	**0.00099**	0.00106	0.00183	0.00202	0.00196	0.00193
COADs	**0.00123**	0.00155	0.00198	0.00231	0.00214	0.00217
Breast tissue	**0.00198**	0.00222	0.00248	0.00293	0.00259	0.00263
Human kidney	**0.00231**	0.00252	0.00259	0.00284	0.00273	0.00274
Gastric	0.00091	0.00122	**0.00061**	.00154	0.00139	0.00137
Lung lobes	0.00011	**0.00008**	0.00138	0.00167	0.00151	0.00157
Lung lesion	**0.00010**	0.00012	0.00534	.00591	0.00552	0.00567
Mice kidneys	**0.00098**	0.00104	0.00137	0.00171	0.00152	0.00160
Mammary gland	0.00017	**0.00011**	0.00266	0.00281	0.00273	0.00269

3.3   Brain Image Registration

For self supervised registration of 3D brain volumes we apply a 3D extension of watershed algorithm, identify structures of interest and apply 3D deformation field. The self supervised registration network is trained to predict the 3D deformation field. $SR - Net$'s difference w.r.t VoxelMorph is the use of self-supervised segmentation maps instead of manually annotated maps. The result is summarized in Table 2. Our proposed registration network, SR-Net, outperforms Voxel Morph and the difference in performance is statistically significant ($p = 0.0013$). The numbers clearly show the significant improvement brought about by self supervised segmentation maps, to the extent that it outperforms manual segmentation maps.

Table 2. Registration results for different methods on brain images.

	Before registration	After registration				
		SR-Net	SR-Net$_{wL_{Seg}}$	DIRNet	FlowNet	VoxelMorph
DM (%)	67.2	**81.2**	77.6	73.0	72.8	79.5
HD$_{95}$ (mm)	14.5	**11.2**	12.6	13.4	13.5	11.9
Time (s)		0.5	0.4	0.6	0.5	0.5

Figure 3 shows results for brain image registration. We show the reference image in Fig. 3(a) followed by an example floating image in Fig. 3(b). The ventricle structure to be aligned is shown in red in both images. Figure 3(c)–(e) shows the deformed structures obtained by applying the registration field obtained from different methods to the floating image and superimposing these structures on the atlas image. The deformed structures from the floating image are shown in blue. In case of perfect registration the blue and red contours should coincide. In this case SR-Net actually does better than VoxelMorph, while SR-Net$_{wL_{Seg}}$ does significantly worse due to absence of segmentation information.

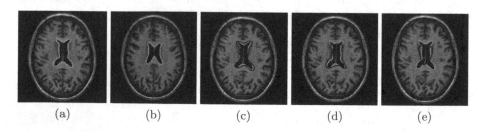

(a)	(b)	(c)	(d)	(e)

Fig. 3. Results for atlas based brain MRI image registration. (a) I_R (b) I_F with manual segmentation in red. Superimposed registered mask (in blue) obtained using: (c) SR-Net; (d) VoxelMorph; (e) SR-Net$_{wL_{Seg}}$. (Color figure online)

4   Conclusion

In this work we have proposed a deep learning based registration method that uses self supervised fine grained segmentation feature maps to include segmentation information in the registration framework. Use of self supervised segmentation maps enables us to include important structural information in scenarios where manual segmentation' maps are unavailable, which is the case for majority of datasets. Experimental results show that by using self supervised segmentation maps the registration results can outperform those obtained using manual segmentation maps. Hence we conclude that self supervised segmentation maps are an effective way of replacing manual segmentation maps and obtaining improved registration. This is applicable for majority of medical image analysis tasks where registration is essential and manual segmentation maps are unavailable.

References

1. ANHIR: Automatic non-rigid histological image registration challenge. https://anhir.grand-challenge.org/. Accessed 30 Jan 2020
2. Bai, W., et al.: Self-supervised learning for cardiac MR image segmentation by anatomical position prediction. In: Shen, D., et al. (eds.) MICCAI 2019. LNCS, vol. 11765, pp. 541–549. Springer, Cham (2019). https://doi.org/10.1007/978-3-030-32245-8_60
3. Balakrishnan, G., Zhao, A., Sabuncu, M.R., Guttag, J., Dalca, A.V.: Voxelmorph: a learning framework for deformable medical image registration. IEEE Trans. Med. Imag. **38**(8), 1788–1800 (2019)
4. Balakrishnan, G., Zhao, A., Sabuncu, M., Guttag, J.: An supervised learning model for deformable medical image registration. In: Proceedings of CVPR, pp. 9252–9260 (2018)
5. Borovec, J., Munoz-Barrutia, A., Kybic, J.: Benchmarking of image registration methods for differently stained histological slides. In: Proceedings of IEEE ICIP, pp. 3368–3372 (2018)
6. Fischl, B.: FreeSurfer. NeuroImage **62**(2), 774–781 (2015)
7. Hu, X., Kang, M., Huang, W., Scott, M.R., Wiest, R., Reyes, M.: Dual-stream pyramid registration network. In: Shen, D., et al. (eds.) MICCAI 2019. LNCS, vol. 11765, pp. 382–390. Springer, Cham (2019). https://doi.org/10.1007/978-3-030-32245-8_43
8. Hu, Y., Gibson, E., Barratt, D.C., Emberton, M., Noble, J.A., Vercauteren, T.: Conditional segmentation in lieu of image registration. In: Shen, D., et al. (eds.) MICCAI 2019. LNCS, vol. 11765, pp. 401–409. Springer, Cham (2019). https://doi.org/10.1007/978-3-030-32245-8_45
9. Jaderberg, M., Simonyan, K., Zisserman, A., Kavukcuoglu, K.: Spatial transformer networks. In: NIPS, pp. 2017–2025 (2015)
10. Kingma, D., Ba, J.: Adam: a method for stochastic optimization. In: International Conference on Learning Representations (2014)
11. Kugler, M., et al.: Robust 3D image reconstruction of pancreatic cancer tumors from histopathological images with different stains and its quantitative performance evaluation. Int. J. Comput. Assist. Radiol. Surg. **14**, 2047–2055 (2019). https://doi.org/10.1007/s11548-019-02019-8

12. Larsson, M., Stenborg, E., Toft, C., Hammarstrand, L., Sattler, T., Kahl, F.: Fine-grained segmentation networks: self-supervised segmentation for improved long-term visual localization. In: Proceedings of ICCV, pp. 31–41 (2019)

13. Lee, M.C.H., Oktay, O., Schuh, A., Schaap, M., Glocker, B.: Image-and-spatial transformer networks for structure-guided image registration. In: Shen, D., et al. (eds.) MICCAI 2019. LNCS, vol. 11765, pp. 337–345. Springer, Cham (2019). https://doi.org/10.1007/978-3-030-32245-8_38

14. Liu, L., Hu, X., Zhu, L., Heng, P.-A.: Probabilistic multilayer regularization network for unsupervised 3D brain image registration. In: Shen, D., et al. (eds.) MICCAI 2019. LNCS, vol. 11765, pp. 346–354. Springer, Cham (2019). https://doi.org/10.1007/978-3-030-32245-8_39

15. Mahapatra, D., Antony, B., Sedai, S., Garnavi, R.: Deformable medical image registration using generative adversarial networks. In: Proceedings of IEEE ISBI, pp. 1449–1453 (2018)

16. Mahapatra, D., Ge, Z.: Training data independent image registration with gans using transfer learning and segmentation information. In: Proceedings of IEEE ISBI, pp. 709–713 (2019)

17. Mahapatra, D., Ge, Z.: Training data independent image registration using generative adversarial networks and domain adaptation. Pattern Recogn. **100**, 1–14 (2020)

18. Mahapatra, D., Ge, Z., Sedai, S., Chakravorty, R.: Joint registration and segmentation of xray images using generative adversarial networks. In: Shi, Y., Suk, H.-I., Liu, M. (eds.) MLMI 2018. LNCS, vol. 11046, pp. 73–80. Springer, Cham (2018). https://doi.org/10.1007/978-3-030-00919-9_9

19. Mahapatra, D., Sun, Y.: Joint Registration and segmentation of dynamic cardiac perfusion images using MRFs. In: Jiang, T., Navab, N., Pluim, J.P.W., Viergever, M.A. (eds.) MICCAI 2010. LNCS, vol. 6361, pp. 493–501. Springer, Heidelberg (2010). https://doi.org/10.1007/978-3-642-15705-9_60

20. Mahapatra, D., Sun, Y.: Integrating segmentation information for improved MRF-based elastic image registration. IEEE Trans. Imag. Proc. **21**(1), 170–183 (2012)

21. Maintz, J., Viergever, M.: A survey of medical image registration. Med. Image Anal. **2**(1), 1–36 (1998)

22. Mueller, S.G.: Ways toward an early diagnosis in Alzheimer's disease: the Alzheimer's disease neuroimaging initiative (ADNI). Alzheimer's Dement. **1**(1), 55–66 (2005)

23. Pathak, D., Krahenbuhl, P., Donahue, J., Darrell, T., Efros, A.A.: Context encoders: feature learning by inpainting. In: Proceedings of CVPR, pp. 31–41 (2016)

24. Pohl, K.M., Fisher, J., Grimson, W.E.L., Kikinis, R., Wells, W.M.: A Bayesian model for joint segmentation and registration. NeuroImage **31**(1), 228–239 (2006)

25. Rohé, M.-M., Datar, M., Heimann, T., Sermesant, M., Pennec, X.: SVF-Net: learning deformable image registration using shape matching. In: Descoteaux, M., Maier-Hein, L., Franz, A., Jannin, P., Collins, D.L., Duchesne, S. (eds.) MICCAI 2017. LNCS, vol. 10433, pp. 266–274. Springer, Cham (2017). https://doi.org/10.1007/978-3-319-66182-7_31

26. Rueckert, D., Sonoda, L., Hayes, C., Hill, D., Leach, M., Hawkes, D.: Nonrigid registration using free-form deformations: application to breast MR images. IEEE Trans. Med. Imaging **18**(8), 712–721 (1999)

27. Sokooti, H., de Vos, B., Berendsen, F., Lelieveldt, B.P.F., Išgum, I., Staring, M.: Nonrigid image registration using multi-scale 3D convolutional neural networks. In: Descoteaux, M., Maier-Hein, L., Franz, A., Jannin, P., Collins, D.L., Duchesne, S. (eds.) MICCAI 2017. LNCS, vol. 10433, pp. 232–239. Springer, Cham (2017). https://doi.org/10.1007/978-3-319-66182-7_27

28. Tibshirani, R., Walther, G., Hastie, T.: Estimating the number of clusters in a data set via the gap statistic. J. Roy. Stat. Soc. B **63**(2), 411–423 (2001)

29. de Vos, B.D., Berendsen, F.F., Viergever, M.A., Staring, M., Išgum, I.: End-to-end unsupervised deformable image registration with a convolutional neural network. In: Cardoso, M., et al. (eds.) DLMIA/ML-CDS -2017. LNCS, vol. 10553, pp. 204–212. Springer, Cham (2017). https://doi.org/10.1007/978-3-319-67558-9_24

30. Xu, Z., Niethammer, M.: DeepAtlas: joint semi-supervised learning of image registration and segmentation. In: Shen, D., et al. (eds.) MICCAI 2019. LNCS, vol. 11765, pp. 420–429. Springer, Cham (2019). https://doi.org/10.1007/978-3-030-32245-8_47

31. Yezzi, A., Zollei, L., Kapur, T.: A variational framework for joint segmentation and registration. In: Proceedings of MMBIA, pp. 44–51 (2001)

Cross-Modality Segmentation by Self-supervised Semantic Alignment in Disentangled Content Space

Junlin Yang[1(✉)], Xiaoxiao Li[1], Daniel Pak[1], Nicha C. Dvornek[3],
Julius Chapiro[3], MingDe Lin[3], and James S. Duncan[1,2,3,4]

[1] Department of Biomedical Engineering, Yale University, New Haven, CT, USA
`junlin.yang@yale.edu`
[2] Department of Electrical Engineering, Yale University, New Haven, CT, USA
[3] Department of Radiology and Biomedical Imaging, Yale School of Medicine,
New Haven, CT, USA
[4] Department of Statistics and Data Science, Yale University, New Haven, CT, USA

Abstract. Deep convolutional networks have demonstrated state-of-the-art performance in a variety of medical image tasks, including segmentation. Taking advantage of images from different modalities has great clinical benefits. However, the generalization ability of deep networks on different modalities is challenging due to domain shift. In this work, we investigate the challenging unsupervised domain adaptation problem of cross-modality medical image segmentation. Cross-modality domain shift can be viewed as having two orthogonal components: appearance (modality) shift and content (anatomy) shift. Previous works using the popular adversarial training strategy emphasize the significant appearance/modality alignment caused by different physical principles while ignoring the content/anatomy alignment, which can be harmful for the downstream segmentation task. Here, we design a cross-modality segmentation pipeline, where self-supervision is introduced to achieve further semantic alignment specifically on the disentangled content space. In the self-supervision branch, in addition to rotation prediction, we also propose elastic transformation prediction as a new pretext task. We validate our model on cross-modality liver segmentation from CT to MR. Both quantitative and qualitative experimental results demonstrate that further semantic alignment through self-supervision can improve segmentation performance significantly, making the learned model more robust.

Keywords: Cross modality · Self supervision · Domain adaptation

1 Introduction

Deep convolutional neural networks have achieved impressive performance on various visual tasks. However, in real life applications, the model performance

X. Li, D. Pak and N.C. Dvornek—Equal Contributions.

© Springer Nature Switzerland AG 2020
S. Albarqouni et al. (Eds.): DART 2020/DCL 2020, LNCS 12444, pp. 52–61, 2020.
https://doi.org/10.1007/978-3-030-60548-3_6

on testing data often degrades heavily due to training and testing data coming from different distributions, referred to as *domain shift* [13]. In particular for medical imaging, each imaging modality is characterized by a different data distribution. The various different modalities with different physical principles unique to medical imaging result in severe domain shifts [5]. Figure 1 shows widely varying pixel value distributions of the same organ in different medical imaging modalities. Bones appear brighter and have greater image contrast in CT than MR, whereas MR has better soft tissue contrast than CT. Other causes of domain shift can include different medical sites, different image acquisition protocols and different scanner manufacturers/models/versions.

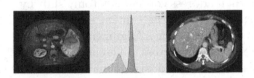

Fig. 1. Different imaging modalities MR and CT with example images and intensity histograms.

Fig. 2. Illustration of our view on the decomposition of domain shift.

It is impractical to tackle the domain shift problem by collecting data and annotations for every new task and new domain. This is especially the case for medical imaging tasks, because image collection and annotation are very expensive, particularly for medical image segmentation. Due to the limited sample size of labeled images for a single modality, it seems to be a natural choice to train a model using one domain, and then, test on another domain. *Unsupervised domain adaptation* serves as a good tool to tackle the domain shift problem and improve the generalizability of learned models on new unseen testing data.

In this work, we focus on the cross-modality medical image segmentation task with unsupervised domain adaptation. We view the domain shift across different modalities as being composed of two orthogonal components: appearance/modality shift and content/anatomy shift.

Adversarial training has been very useful in a series of medical imaging applications [3,5,10,14], including cross-modality medical image analysis. Current popular works on cross-modality medical image analysis use similar adversarial training strategies [2,3,5,7,14]. Most require a discriminator to differentiate between two different modalities on image-level or feature-level. As the discriminator can be easily fooled, domain adaptation via adversarial training is unstable and biased to only eliminate appearance/modality shift while ignoring the content/anatomy shift. In other words, it doesn't necessarily achieve semantic alignment, which is important for downstream tasks.

Self-supervised learning uses automatically labeled training data to eliminate the cost and mitigate the lack of manual annotations. It is still supervised learning, but it can get supervision from the data itself. The learning tasks in self-supervision are called pretext tasks. For example, we might randomly rotate images as inputs and the pretext task is to predict how each image is rotated. It has been used for semi-supervised learning [15], domain generalization [1], etc. By learning pretext tasks, the model will have a better semantic or structural understanding of the data, resulting in better downstream task performance.

Here we argue that it is important to semantically align content/anatomy across modalities, besides tackling the appearance/modality shift, as shown in Fig. 2. We designed a cross-modality segmentation model with disentangled representations to embed images into appearance space and content space. Disentangled learning is supervised by adversarial training to eliminate the appearance/modality shift in content space. We then address content/anatomy shift by introducing a self-supervision strategy specifically in the content space, which provides a new way of encouraging semantic alignment. Alignment is encouraged through shared self-supervised pretext tasks across different modalities instead of adversarial training between generators and discriminators. Shared self-supervised tasks encourage the model to align content space from different domains semantically. The main contributions are summarized as follows:

- This work addresses the neglected content/anatomy alignment problem while solving the significant appearance/modality shift in the cross-modality segmentation setting. The proposed unique pipeline directly estimates the segmentation from the content space encoding and utilizes self-supervision to achieve further semantic alignment in the disentangled content space.
- The proposed model has been validated on cross-modality liver segmentation from CT to MR and shown significant improvements.
- As an early effort to incorporate self-supervision for the cross-modality segmentation problem, it demonstrates the effectiveness of self-supervised semantic alignment in content space. Besides rotation prediction, a new self-supervision pretext task of elastic transformation prediction is proposed.

2   Methodology

2.1   Problem Formulation

Given source data x^S with labels and target data x^T without labels, unsupervised domain adaptation aims to learn a model from source data that works on target data. In the setting of cross-modality medical image segmentation, source data and target data are unpaired data from different modalities, with a shared segmentation task on objects of interest. It is believed that domain shift between different modalities are relatively large due to different physical principles. We argue that domain shift across different modalities is composed of two orthogonal components, appearance/modality shift (obvious) and content/anatomy shift (often ignored). The commonly used adversarial training strategy is effective at reducing appearance/modality shift, but not as effective or sometimes even worsens content/anatomy shift. One example of this would be a fake MR image generated from a CT image may look exactly like an MR image in terms of intensity distributions, but may have organs that do not match the original CT image in terms of location or shape.

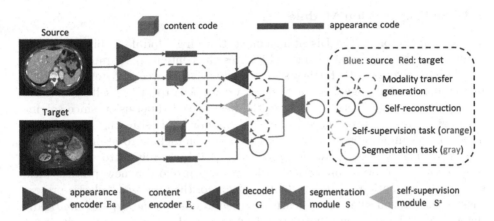

Fig. 3. Overview of our proposed domain adaptation pipeline by self-supervised semantic alignment in disentangled content space.

2.2 Overall Framework

To further handle the content/anatomy shift, we propose semantic alignment in content space using the framework shown in Fig. 3. It is composed of three parts. (1) The disentangled representation learning module embeds images into content and appearance spaces through a modality transfer generation path and generates content-only images (generated using content encoding with appearance code set to 0 vector). (2) The self-supervision module achieves further semantic alignment in content space through self-supervised pretext tasks. (3) The segmentation module makes predictions on content-only images from both source and target modalities.

The three parts of the model are jointly trained end-to-end, where the disentanglement learning module and self-supervision module tackle the appearance shift and achieve content alignment semantically for the downstream segmentation task, respectively. Details for each module are presented below.

2.3 Disentanglement Learning Module

To eliminate the appearance shift component of domain shift, similar to [9,14], disentangled representations are learned to disentangle content and appearance of images via supervision by adversarial training. It takes the images from different domains as inputs and outputs content-only images without domain information. It is believed that appearance shift between two different modalities has been eliminated in the content space and content-only images. Different from the two stage model in [14], the Disentanglement Learning Module is trained end-to-end with the Segmentation Module.

2.4 Self-supervision Module

The discriminator in the Disentanglement Learning Module is biased toward eliminating the appearance shift and ignores the content misalignment, thus it can be easily fooled. To further semantically align source and target domain in the above disentangled content space, we propose to utilize self-supervision with pretext tasks. Pretext tasks are the key for self-supervision since solving different pretext tasks requires different semantic understanding of the data. The most common pretext tasks for self-supervision include rotation prediction, flip prediction and patch location prediction [1,6]. In addition to adopting the most common rotation prediction task, we also propose a new pretext task, elastic transformation prediction. Our intuition for the proposed +elastic pretext task is that it will allow us to learn representations for the type of nonrigidly deformable objects/shapes that should be seen in abdominal images, rather than for arbitrary elastic transformations. In our model, the content space encodings from different domains are fed into the self-supervision head directly to predict the self-supervised tasks.

Fig. 4. Example rotated MR images for the rotation pretext task. Left to right: 0, 90, 180, 270°.

Fig. 5. Example CT and MR original images and elastic transformed images for the elastic pretext task.

For rotation prediction, as shown in Fig. 4, input images from the two modalities are rotated randomly by 90, 180 and 270° and fed into content encoders. The self-supervision head is a 4-class classifier that takes the content codes as input and predicts the angle of rotation. For the proposed elastic transformation prediction, as shown in Fig. 5, elastic transformations as described in [12] are randomly applied to 50% of the input images from the two modalities. The task is to predict whether the input images have elastic transformations.

We hypothesize that accurate rotation prediction and elastic transformation prediction requires the content space encoding to preserve certain semantic information that is beneficial for downstream tasks. Furthermore, the pretext tasks are learned on two modalities together using the same self-supervision head; this should encourage semantic alignment in the content space across two domains.

2.5 Segmentation Module

The segmentation module handles the task of interest. It takes content-only images generated by the semantically aligned content space encoding in the disentanglement learning module as input and outputs the segmentation prediction.

2.6 Implementation

Disentanglement learning module consists of two content encoders E_c^S, E_c^T and two appearance encoders E_a^S, E_a^T. Images are embedded into two different spaces, content space and appearance space. S refers to source modality and T refers to target modality. Content code is a feature map, denoted as $c^S, c^T \in \mathbb{R}^{n \times n \times k}$. Appearance code is a vector, denoted as $a^S, a^T \in \mathbb{R}^m$. Two decoders G^S, G^T with Adaptive Instance Normalization (AdaIN) [8] take the content and style codes to achieve self-reconstruction and modality transfer generation. Two discriminators D^S, D^T that differentiate between two modalities are used to compete with decoders for modality transfer generation in an adversarial training way.

For self-reconstruction, content and appearance codes from the same modality are fed into decoders, and L1 loss is computed in the image domain as shown in equation (1). Of note, $c^i = E_c^i(x^i)$ and $a^i = E_a^i(x^i)$ for $i = S$ or T.

$$L_{recon} = \mathbb{E}||G^S(c^S, a^S) - x^S||_1 + \mathbb{E}||G^T(c^T, a^T) - x^T||_1 \tag{1}$$

The modality transfer generation loop supervises the disentanglement learning in an adversarial training way. Content and appearance codes from different modalities are fed into decoders G^S, G^T. Two discriminators compete with the modality transfer generation module, thus encouraging the disentanglement of content and appearance. Please refer to Eq. (2). Of note, $x^{T \to S} = G^S(c^T, a^S)$, $x^{S \to T} = G^T(c^S, a^T)$.

$$L_{adv}^{S \to T} + L_{adv}^{T \to S} = \mathbb{E}[log(1 - D^S(x^{T \to S}))] + \mathbb{E}[log(D^T(x^T))]$$
$$+ \mathbb{E}[log(1 - D^T(x^{S \to T}))] + \mathbb{E}[log(D^S(x^S))] \tag{2}$$

In addition, Cycle-consistency loss on content and style codes is enforced and the appearance code is trained to match the standard normal distribution by minimizing the Kullback-Leibler (KL) divergence.

The self-supervision pretext task is jointly trained with the main task of interest. In our case, cross entropy loss is calculated to update the self-supervision head H and content encoders E_c^S, E_c^T.

Setting style codes to zero, decoders G^S, G^T take only content codes and output content-only images. Content-only images serve as the inputs of the Segmentation Module and cross entropy loss is calculated to train the model.

For validation, we must assume there is no ground truth for the target modality to compute cross entropy as validation loss. In practice, we computed the distance in content feature space on validation data between source and target as our validation loss to determine the stopping criterion.

3   Experiments

3.1   Datasets and Evaluation Metric

Unpaired 131 contrast-enhanced CT scans from LiTS 2017 MICCAI [4] and 55 T1 dynamic contrast enhanced MR scans in arterial phase from local hospitals are utilized for the experiments. We resampled and sliced both CT and MR into 256×256 in the transverse plane, uniformly sampling $256 \times 256 \times 3$ slices from each patient as inputs. CT is much cheaper and more available than MR scans and we are interested in achieving liver segmentation with domain adaptation from public labeled CT to private unlabeled MR. We evaluate the segmentation performance with the comprehensive metrics of volume Dice coefficient, average symmetric surface distance and robust Hausdorff distance by calculating the subject average and standard deviation of the segmentation results.

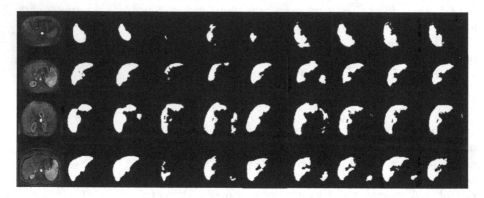

Fig. 6. Examples of liver segmentation results. From left to right: MR, Ground Truth, Oracle, UNet, CycleGAN, SIFA, DADR, Disentangled, + Elastic, + Rotation

3.2   Experiment Settings

We tested our model on the cross-modality liver segmentation task. In our experiments, data from each modality are randomly split patient-wise with 70% volumes for training, 10% volumes for validation and 20% volumes for testing.

We individually tested our disentangled model without self-supervision, disentangled model with elastic transformation prediction as self-supervision, and disentangled model with rotation prediction as self-supervision, denoted as Disentangled, + Elastic, + Rotation, respectively, in Table 1.

For comparison, we tested popular unsupervised domain adaptation methods, CycleGAN [16], SIFA [3] and DADR [14] on the same datasets. The CycleGAN and DADR approaches are both two-step processes which learn the mapping between source and target domains. Then, a UNet [11] is trained using the labeled source images transferred to the target domain and used to predict test data in the target domain. SIFA is a one stage model that aligns style at both image and feature level.

Table 1. Liver segmentation results, Dice scores (DSC, %) and average symmetric surface distance (ASSD, mm), robust Hausdorff distance (HD, mm)

Method	DSC avg. ± std	ASSD avg. ± std	HD avg. ± std
Oracle	92.92 ± 1.27	0.96 ± 1.66	2.75 ± 0.78
UNet [11]	52.65 ± 17.59	4.74 ± 4.32	16.40 ± 11.71
CycleGAN [16]	79.41 ± 4.96	2.42 ± 1.73	5.92 ± 5.41
SIFA [3]	82.62 ± 2.31	2.37 ± 1.68	5.46 ± 4.71
DADR [14]	83.23 ± 3.45	2.82 ± 3.25	6.38 ± 6.43
Disentangled	84.18 ± 2.94	2.43 ±1.78	5.87 ± 3.20
+ Elastic	**86.48 ± 2.53**	2.39 ± 1.76	5.28 ± 3.43
+ Rotation	86.17 ± 2.28	**2.24 ± 1.89**	**4.47 ± 2.96**

3.3 Results and Analysis

Results for experiments on the liver segmentation with domain adaptation are shown in Table 1. The Oracle result is from training labeled data in the target MR domain. UNet trained on the source CT domain without adaptation to the target MR domain performed worst, as expected. Domain adaptation using CycleGAN, SIFA and DADR models produced improved segmentation results. Our end-to-end baseline Disentangled model resulted in higher DSC, ASSD and HD compared to other methods that do not use self-supervision. The proposed models with self-supervision + Elastic and + Rotation further increased segmentation accuracy, with the two pretext tasks resulting in comparable segmentation performance. Example segmentation results are shown in Fig. 6.

Both elastic transformation prediction and rotation prediction improve from the baseline disentangled model, which shows the effectiveness of the proposed self-supervised semantic alignment in content space. Although the quantitative results are comparable for + Elastic and + Rotation, the qualitative results of + Elastic and + Rotation in Fig. 6 showed visual differences on improved segmentation results. We observed that different self-supervised tasks require different semantic information, thus providing semantic alignment in different directions.

4   Conclusions

In this paper, we tackled the clinically important cross-modality segmentation problem. In particular, we view the domain shift as two orthogonal components, appearance/modality and content/anatomy shift. Upon handling the significant appearance shift through disentanglement learning, we proposed to achieve further semantic alignment in content space through self-supervised tasks. Through experiments on cross-modality liver segmentation, we showed the effectiveness of semantic alignment introduced by self-supervision on cross-modality segmentation. We observed that different self-supervised tasks require different semantic information, thus providing semantic alignment in different directions. For future work, we recommend achieving semantic alignment through self-supervision. The key to self-supervision is to design suitable self-supervised tasks and determine how to combine different self-supervision strategies to provide synergy for solving the target task.

Acknowledgement. This work was supported by NIH Grant 5R01 CA206180

References

1. Carlucci, F.M., D'Innocente, A., Bucci, S., Caputo, B., Tommasi, T.: Domain generalization by solving jigsaw puzzles. In: Proceedings of the IEEE Conference on Computer Vision and Pattern Recognition, pp. 2229–2238 (2019)
2. Chartsias, A., et al.: Disentangled representation learning in cardiac image analysis. Med. Image Anal. **58**, 101535 (2019)
3. Chen, C., Dou, Q., Chen, H., Qin, J., Heng, P.A.: Synergistic image and feature adaptation: towards cross-modality domain adaptation for medical image segmentation. In: Proceedings of the AAAI Conference on Artificial Intelligence, vol. 33, pp. 865–872 (2019)
4. Christ, P., Ettlinger, F., Grün, F., Lipkova, J., Kaissis, G.: LiTS-liver tumor segmentation challenge. ISBI and MICCAI (2017)
5. Dou, Q., et al.: Pnp-adanet: plug-and-play adversarial domain adaptation network at unpaired cross-modality cardiac segmentation. IEEE Access **7**, 99065–99076 (2019)
6. Gidaris, S., Singh, P., Komodakis, N.: Unsupervised representation learning by predicting image rotations. In: International Conference on Learning Representations (2018). https://openreview.net/forum?id=S1v4N2l0-
7. Hoffman, J., et al.: Cycada: cycle-consistent adversarial domain adaptation. arXiv preprint arXiv:1711.03213 (2017)
8. Huang, X., Belongie, S.: Arbitrary style transfer in real-time with adaptive instance normalization. In: Proceedings of the IEEE International Conference on Computer Vision, pp. 1501–1510 (2017)
9. Huang, X., Liu, M.Y., Belongie, S., Kautz, J.: Multimodal unsupervised image-to-image translation. In: Proceedings of the European Conference on Computer Vision (ECCV), pp. 172–189 (2018)
10. Li, W., Wang, Y., Cai, Y., Arnold, C., Zhao, E., Yuan, Y.: Semi-supervised rare disease detection using generative adversarial network. arXiv preprint arXiv:1812.00547 (2018)
11. Ronneberger, O., Fischer, P., Brox, T.: U-net: convolutional networks for biomedical image segmentation. In: Navab, N., Hornegger, J., Wells, W.M., Frangi, A.F. (eds.) MICCAI 2015. LNCS, vol. 9351, pp. 234–241. Springer, Cham (2015). https://doi.org/10.1007/978-3-319-24574-4_28

12. Simard, P.Y., Steinkraus, D., Platt, J.C., et al.: Best practices for convolutional neural networks applied to visual document analysis. In: ICDAR, pp. 958–962 (2003)
13. Wang, M., Deng, W.: Deep visual domain adaptation: a survey. Neurocomputing **312**, 135–153 (2018)
14. Yang, J., Dvornek, N.C., Zhang, F., Chapiro, J., Lin, M.D., Duncan, J.S.: Unsupervised domain adaptation via disentangled representations: application to cross-modality liver segmentation. In: Shen, D., et al. (eds.) MICCAI 2019. LNCS, vol. 11765, pp. 255–263. Springer, Cham (2019). https://doi.org/10.1007/978-3-030-32245-8_29
15. Zhai, X., Oliver, A., Kolesnikov, A., Beyer, L.: S4l: self-supervised semi-supervised learning. arXiv preprint arXiv:1905.03670 (2019)
16. Zhu, J.Y., Park, T., Isola, P., Efros, A.A.: Unpaired image-to-image translation using cycle-consistent adversarial networks. In: Proceedings of the IEEE International Conference on Computer Vision, pp. 2223–2232 (2017)

Semi-supervised Pathology Segmentation with Disentangled Representations

Haochuan Jiang[7]([✉]), Agisilaos Chartsias[1], Xinheng Zhang[2,3],
Giorgos Papanastasiou[4], Scott Semple[5], Mark Dweck[5], David Semple[5],
Rohan Dharmakumar[3,6], and Sotirios A. Tsaftaris[1]

[1] School of Engineering, University of Edinburgh, Edinburgh, UK
[2] Department of Bioengineering, University of California, Berkeley, USA
[3] Biomedical Imaging Research Institute, Cedars-Sinai Medical Center,
Los Angeles, USA
[4] School of Computer Science and Electronic Engineering, University of Essex,
Colchester, UK
[5] Center for Cardiovascular Science, University of Edinburgh, Edinburgh, UK
[6] Department of Medicine, University of California, Berkeley, USA
[7] School of Robotics, Xi'an Jiaotong-Liverpool University, Suzhou, P.R. China
haochuan.jiang@xjtlu.edu.cn

Abstract. Automated pathology segmentation remains a valuable diagnostic tool in clinical practice. However, collecting training data is challenging. Semi-supervised approaches by combining labelled and unlabelled data can offer a solution to data scarcity. An approach to semi-supervised learning relies on reconstruction objectives (as self-supervision objectives) that learns in a joint fashion suitable representations for the task. Here, we propose Anatomy-Pathology Disentanglement Network (APD-Net), a pathology segmentation model that attempts to learn jointly for the first time: disentanglement of anatomy, modality, and pathology. The model is trained in a semi-supervised fashion with new reconstruction losses directly aiming to improve pathology segmentation with limited annotations. In addition, a joint optimization strategy is proposed to fully take advantage of the available annotations. We evaluate our methods with two private cardiac infarction segmentation datasets with LGE-MRI scans. APD-Net can perform pathology segmentation with few annotations, maintain performance with different amounts of supervision, and outperform related deep learning methods.

Keywords: Pathology segmentation · Disentangled representations · Semi-supervised learning

1 Introduction

Deep learning models for automated segmentation of pathological regions from medical images can provide valuable assistance to clinicians. However, such mod-

---

H. Jiang — Most of the work was completed in the School of Engineering at the University of Edinburgh

S. Albarqouni et al. (Eds.): DART 2020/DCL 2020, LNCS 12444, pp. 62–72, 2020.
https://doi.org/10.1007/978-3-030-60548-3_7

els require a considerable amount of annotated data to train, which may not be easy to obtain. Pathology annotation (as opposed to anatomical), relies also on carefully detecting normal tissue areas for direct comparison. It is therefore appealing to train pathology segmentors by combining the available annotated data with larger numbers of unlabeled images in a semi-supervised learning scheme.

A typical strategy to segment pathology is to first locate the affected anatomy, e.g. the myocardium for cardiac infarction [9], and use the detected anatomy to guide the pathology prediction. However, since anatomical annotations are not always available, recent methods cascade two networks: one segments the anatomy of interest, and the second one segments the pathology [10,11,14].

Although these methods achieve accurate segmentation, they are typically fully supervised and sensitive annotations numbers, as seen in our experimental results specified in Sec. 4. Semi-supervised learning is promising to solve the issue by engaging unlabelled images. Recently, disentangled representations, i.e. structured latent spaces that are shared between labeled and unlabeled data, have provided a solution to semi-supervised learning [3,6]. These methods typically use specialized encoders to separate anatomy (a spatial tensor) and imaging information (a vector encoding image appearance) in medical image applications [3], while they involve unlabeled data through reconstruction losses.

In this paper, inspired by [3], we propose the Anatomy-Pathology Disentanglement Network (APD-Net). APD-Net constructs a space of anatomy (a spatial tensor), modality (a vector), and pathology (a spatial tensor) latent factors. Pathology is obtained by an encoder, which segments the pathology conditioned on both the image and the predicted anatomy mask. We focus on segmenting myocardial infarct, a challenging task due to its size, irregular shape and random location. APD-Net is optimized with several objectives. Among others, we introduce a novel ratio-based triplet loss to encourage the reconstruction of the pathology region by taking advantage of the pathology factor. In addition, the use of reconstruction losses, that are made possible with disentangled representations [3], makes APD-Net suitable for semi-supervised learning. Finally, we train with both predicted and real anatomy and pathology masks (in a *Teacher-Forcing* strategy [19]) to further improve performance. Our major **contributions** are summarized as follows:

- We propose a method for disentangled representations of anatomy, modality, and pathology;
- The disentanglement is encouraged with a novel ratio-based triplet loss;
- We also proposed the *Teacher-Forcing* training regime combining different scenarios of real and predicted inputs to make full use of available annotations during optimization;
- APD-Net improves the Dice score of state-of-the-art benchmarks on two private datasets for cardiac infarction segmentation when limited supervision is present, whilst maintaining performance in the full annotation setting.

2 Related Work

Pathology Segmentation: A classical approach to segment pathology is by cascading organ segmentation before segmenting the pathological region. These two segmentors can be trained separately [10], or jointly [11,14]. In contrast to anatomy, pathology is small in sizes and irregular in shapes; thus, shape priors cannot be used. To train models when masks are small, the Tversky loss [14,15] or similarly the focal loss in [1] have been proposed. Our proposed model also cascades two segmentations, by using the initial anatomy prediction to guide the subsequent pathology segmentation. However, we achieve this using disentangled representations to enable semi-supervised learning.

Disentangled Representations. The idea of disentangled representations is to decompose the latent space into domain-invariant spatial content latent factors (known as anatomy in medical imaging) and domain-related vector style ones (here referred to as modality) [6,12]. In medical image analysis, images are disentangled in anatomical and imaging factors for the purpose of semi-supervised segmentation [3], image registration [12], and classification [18]. Image reconstruction for semi-supervised segmentation was also investigated in [5] with a simpler disentanglement of the foreground (predicted anatomy masks) and the remaining background. However, less effort has been placed on disentangling pathology. Some pioneering studies were conducted in [20], treating brain lesion segmentations as a pathology factor to synthesise pseudo-healthy images. In this paper, we also adopt a segmentation of pathology as a latent factor and combine it with disentangled anatomy and modality [3], which enables the image reconstruction task for semi-supervised learning. While we are inspired by others [20] who consider anatomy and pathology factors independently. This work is the first to learn them in a joint fashion.

3 Methodology

This section presents the APD-Net model. We first introduce relevant notations. Then, we specify disentanglement properties (Sec. 3.1), detail the model architecture (Sec. 3.2), and finally present the learning objectives (Sec. 3.3) with the joint training strategy depending on the different input scenarios (Sec. 3.4).

Notation: Let X, Y_{ana}, Y_{pat} be sets of volume slices, and the associated anatomy and pathology masks, respectively. Let i be a sample. We assume a fully labeled pathology subset $\{x^i, y_{ana}^i, y_{pat}^i\}$, where $x^i \in X \subset \mathbb{R}^{H \times W \times 1}$, $y_{ana}^i \in Y_{ana} := \{0,1\}^{H \times W \times N}$, and $y_{pat}^i \in Y_{pat} := \{0,1\}^{H \times W \times K}$. N and K denote the number of anatomy, and pathology masks respectively. H and W are the image height and width. When $Y_{pat} = \emptyset$, it degrades to an unlabeled pathology set.[1] Both anatomy and pathology sets are involved in a semi-

---

[1] We only consider unlabeled pathology, assuming anatomy masks are available during training. Partial anatomy annotation is out of the scope of this paper.

supervised fashion to segment pathology. This is achieved by learning a mapping function f that estimates anatomy and pathology given an image x^i, i.e. $\{\hat{y}_{ana}^i, \hat{y}_{pat}^i\} = f(x^i)$.

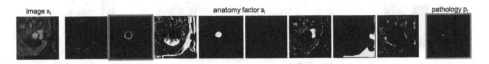

image x_i anatomy factor s_i pathology p_i

Fig. 1. Visualising the disentanglement of the spatial (and binary) anatomy and pathology factors for a LGE-MRI slice. (Color figure online)

3.1 Pathology Disentanglement

The main idea of APD-Net is to take an input image and decompose it into latent factors that relate to anatomy, pathology, and image appearance (modality). This will allow inference of anatomical and pathology segmentations, whereas the disentanglement of the image appearance will enable image reconstruction that is critical to enable training with unlabeled images and semi-supervised learning. Herein we consider $C = 8$ channels of anatomy factors and $n_z = 8$ for modality factors as in [3], and $K = 1$ (myocardial infarct). s^i and p^i are obtained by softmax and sigmoid output activations respectively. They are then binarised (per-channel) by $s^i- > [s^i + 0.5]$ and $p^i- > [p^i + 0.5]$, such that each pixel corresponds to exactly one channel. This binarisation encourages the produced anatomy factor to be modality-invariant. Finally, as in [3] gradients are bypassed in the backward pass to enable back-propagation.

Figure 1 illustrates predicted anatomy and pathology factors for a cardiac infarct example. We make an intuitive distinction between the two factors in that the former only refers to healthy anatomical regions. Therefore, there is an overlap between the pathology factor and one or more anatomy channels. In this example (Fig. 1), the pathology (infarct in the green box) is spatially correlated with the myocardial channel (red box). Encoding pathology in the anatomy factor (i.e. entanglement of these two) is prevented, both through architecture design and with relevant losses. This will be detailed in the following sections.

3.2 APD-Net Architecture

APD-Net, depicted in Fig. 2, adopts modules from SD-Net [3] including anatomy Enc_{ana} and modality Enc_{mod} encoders, anatomy segmentor Seg_{ana}, and decoder Dec. They give s^i, z^i, the anatomy mask $\hat{y}_{ana}^i$, and the reconstructed image $\hat{x}^i$.

We introduce a pathology encoder Enc_{pat} following the U-Net [13] architecture. Given channel-wise concatenated x^i and $\hat{y}_{ana}^i$, Enc_{pat} produces p^i.[2] Thus, APD-Net structurally resembles the cascaded segmentation scheme [11], enabling Enc_{pat} to focus on specific regions to locate the pathological tissue.

---

[2] Note that p^i is the same as $\hat{y}_{pat}^i$, i.e. the predicted pathology mask. We use p^i for disentanglement and image reconstruction, and $\hat{y}_{pat}^i$ for pathology segmentation.

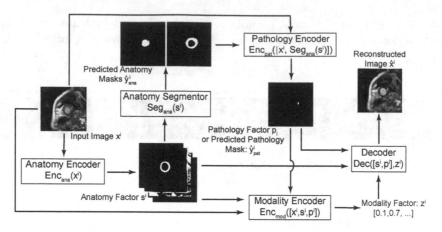

Fig. 2. Schematic of APD-Net. An image is encoded to anatomy factors using Enc_{ana}, and segmented with Seg_{ana} to produce anatomical segmentation masks (in this case the myocardium and left ventricle). Combined with the input, the anatomy segmentation is used to segment the pathology with Enc_{pat}. Finally, given the anatomy, the pathology, and the modality factors from Enc_{Mod}, the decoder reconstructs the input.

Finally, the image decoder receives the concatenation of s^i, p^i, and z^i to reconstruct image $\hat{x}^i$, enabling unsupervised training. With nulled pathology $p_0^i = \mathbb{0}$ (all elements are zero), a pseudo-healthy image ($\hat{x}_0^i$) is obtained [20].

3.3   Individual Training Losses

APD-Net is jointly trained with losses including new supervised, unsupervised, and objectives selected from [3,4] for the task of pathology segmentation.

Pathology Supervised Losses: Pathology manifests in various shapes. Thus, using a mask discriminator as a shape prior [2] is not advised. In addition, pathology covers a small portion of the image leading to class imbalance between the foreground and background. To address these shortcomings, inspired by [14, 17], we combine Tversky[3] [15] and focal loss [8]. The Tversky loss is defined as follows: $\ell_{patT} = (\hat{y}_{pat}^i \odot y_{pat}^i)/[\hat{y}_{pat}^i + y_{pat}^i + (1-\beta) \cdot (\hat{y}_{pat}^i - \hat{y}_{pat}^i \odot y_{pat}^i) + \beta \cdot (y_{pat}^i - \hat{y}_{pat}^i \odot y_{pat}^i)]$, where $\odot$ represents the element-wise multiplication. The focal loss is defined as $\ell_{patF} = \sum_{H,W}[-y_{pat}^i(1 - \hat{y}_{pat}^i)^\gamma \log(\hat{y}_{pat}^i)]$.

Pathology-Masked Image Reconstruction Loss: Typically image reconstruction is achieved by minimising the ℓ_1 or ℓ_2 loss between x^i and $\hat{x}^i$. However, due to the size imbalance between pathological and healthy regions, conventional full image reconstruction may ignore (average out) the pathology region. A simple, but effective solution, is to measure reconstruction performance on the pathology region by using the real pathology mask. This is implemented by a

---

[3] The Dice loss can be seen as a special case of the Tversky loss [15] when $\beta = 0.5$.

masked reconstruction ℓ_1 loss: $\ell_M = \frac{1}{H \times W \times C} \sum_{H,W,C} (\frac{\lambda_{pat}}{\lambda_{ana}} \cdot y^i_{pat} + \mathbb{1}) \cdot \|x^i - \hat{x}^i\|_1$. $\mathbb{1}$ denotes a matrix with y^i_{pat} dimensions, where all elements are ones.[4]

Ratio-based Triplet Loss: However, this may not be adequate for accurate pathology reconstruction. We therefore penalise the model when pathology is possibly ignored by adopting a contrastive Triplet loss [16]. This is defined as $\max(m + d_{pos} - d_{neg}, 0)$, and minimizes the inter-class distance (d_{pos}) compared to the intra-class (d_{neg}) in deep feature space based on a margin (m). We generate *pseudo-healthy* images as negative examples, $\hat{x}^i_0 = Dec(s^i, p^i_0, z^i)$, obtained by nulling the pathology factor, i.e. p^i_0. The deep features are calculated as a new output attached to the penultimate layer of a reconstruction discriminator (inherent from the SD-Net [4] for adversarial loss on $\hat{x}^i$) for the decoder output (denoted as T). By choosing $T(\hat{x}^i)$ as the anchor [16], positive and negative distances are calculated as $d_{pos} = \|T(\hat{x}^i) - T(x^i)\|^2_2$ and $d_{neg} = \|T(\hat{x}^i) - T(\hat{x}^i_0)\|^2_2$ respectively with corresponding samples $T(x^i)$ and $T(\hat{x}^i_0)$.

In practice, choosing a proper m value is challenging [21], particularly when the difference between the positive and the negative samples only lies in the small pathology region, e.g. in the cardiac infarction of Fig. 1. This will lead to an extremely small difference. Instead of optimizing the absolute margin m, we propose to alternatively minimize the Ratio-based triple loss (RT) defined as $\ell_{RT} = \max(r + \frac{d_{pos}}{d_{neg}} - 1, 0)$. The hyper-parameter r represents the relative margin that the positive should be closer to the anchor than the negative.

SD-Net Losses: We adopt optimization objectives from SD-Net [3] and its multi-modal extension [4] to the proposed framework. Specifically, the anatomy supervision Dice loss, the modality factor KL divergence, and the factor reconstruction loss are inherent from [3], while the adversarial loss on $\hat{x}^i$ is brought from [4]. We refer SD-Net losses as ℓ_{SDNet}.

3.4 Joint Optimization with *Teacher-Forcing* Training Strategy

A critical issue in cascaded architectures is that initial segmentation errors propagate and directly affect the second prediction. This should be taken into account particularly during training. We thus adopt the *Teaching-Forcing* strategy [19], originally applied in RNNs by engaging real rather than predicted labels.

In APD-Net, this strategy is applied on pathology segmentation $\hat{y}^i_{pat}$, modality factor estimation z^i, and reconstruction $\hat{x}_i$, which depend on real or predicted anatomy and pathology segmentations. Specifically, $\hat{y}^i_{pat}$ can be estimated using predicted anatomy masks, $\hat{y}^i_{pat} = Enc_{pat}([x^i, Seg_{ana}(Enc_{ana}(x^i))])$, or real masks $\hat{y}^i_{pat} = Enc_{pat}([x^i, y^i_{ana}])$. Subsequently, the modality factor can be produced by the predicted pathology mask, $z^i = Enc_{mod}([x^i, Enc_{ana}(x^i), \hat{y}^i_{pat}])$, or

---

[4] The $\mathbb{1}$ matrix is added to ensure that no zero elements are multiplied with $\|x^i - \hat{x}^i\|_1$. Also, if $\lambda_{pat} = 1$, the loss reduces to the ℓ_1 loss.

the real pathology mask, $z^i = Enc_{mod}([x^i, Enc_{ana}(x^i), y^i_{pat}])$. Finally, real or predicted pathology contributes to reconstruction, $\hat{x}^i = Dec([Enc_{ana}(x^i), y^i_{pat}], z^i)$ and $\hat{x}^i = Dec([Enc_{ana}(x^i), \hat{y}^i_{pat}], z^i)$, respectively.

We term the losses that involve the predicted, real anatomy mask, and real pathology mask as ℓ^{PA}, ℓ^{RA}, and ℓ^{RP} respectively. We redefine each loss as a weighted sum $\ell = \lambda^{PA}\ell^{PA} + \lambda^{RA}\ell^{RA} + \lambda^{RP}\ell^{RP}$, where λ^{PA}, λ^{RA}, and λ^{RP} are relative weights, and $\ell \in \{\ell_z, \ell_{RT}, \ell_{adv}, \ell_M, \ell_{KL}\}$. Finally, the full objective is given by $\ell_{APD-Net} = \lambda_{patT}\ell_{patT} + \lambda_{patF}\ell_{patF} + \lambda_{RT}\ell_{RT} + \lambda_M\ell_M + \ell_{SDNet}$.

4   Experiments

We evaluate APD-Net on pathology segmentation using the Dice score. Experimental setup, datasets, benchmarks, and training details will be detailed below.

Data: We use two private cardiac LGE datasets acquired at the Biomedical Imaging Research Institute of the Cedars-Sinai Medical Center (*Data1*) and the Center for Cardiovascular Science of the University of Edinburgh (*Data2*), which have been approved by data ethics committees of the respective providers. Both datasets contain annotations of the myocardium and myocardial infarct. *Data1* involves 45 subjects (36 used for training) and 224×224 dimension. *Data2* consists of 26 (mixed healthy and pathology) subjects (20 used for training), and 192×192 dimension.

Benchmarks: We compare APD-Net with three benchmarks on infarct segmentation. *U-Net (masked):* A U-Net trained on images x^i masked by the ground truth myocardium mask y^i_{ana} [9]. Masking here facilitates training, reducing the task to finding only infarcted myocardial pixels. *U-Net (unmasked):* The U-Net is trained on images x_i without masking. This is more challenging since now the U-Net implicitly has to find infarct pixels from the whole image. *Cascaded U-Net* [11]: this trains one U-Net to segment the myocardium (with 100% supervision) and another to segment the infarct after masking the input image with the predicted myocardium (varying the number of available annotations).[5]

Training Details: Training penalties are set to: $\lambda_{patT}=1$, $\lambda_{patF}=1.5$, $\lambda_{RT}=1$, $\lambda_M=3$. Weights for different optimization scenarios are: $\lambda^{PA}=1$, $\lambda^{RA}=0.7$, and $\lambda^{RP}=0.5$. The relative margin for ℓ_{RT} is $r=0.3$. Other hyper-parameters include $\beta=0.7$ in ℓ_{patT}, $\gamma=2$ in ℓ_{patF}, and the SD-Net weights defined in [3]. Due to the small data size, we do not specify validation sets. All models are trained for fixed 100 epochs, and results reported below contain averaged Dice scores and standard deviation on test data of two different splits.[6]

---

[5] *U-Net (masked / unmasked)* and *Cascaded U-Net* are optimized with full supervision using Tversky and focal losses, and penalized as defined in the **Training details**. In reality, *U-Net (masked)* is not a good choice since manual myocardial annotations are not always available at inference time.

[6] Code will be available at https://github.com/falconjhc/APD-Net shortly.

Table 1. Performance evaluation for *Data1* and *Data2*. We report test Dice scores (with standard deviation in subscript calculated by summarizing all the involved volumes) on infarct segmentation with varying infarct supervision (% infarct).

Dataset - % infarct	U-Net (unmasked)	Cascaded U-Net	APD-Net	U-Net (masked)
Data1-13%	$5.4_{8.1}$	$36.2_{17.9}$	$45.3_{14.4}$	$57.3_{13.5}$
Data1-25%	$6.5_{8.4}$	$39.4_{15.6}$	$46.4_{12.8}$	$57.3_{13.2}$
Data1-50%	$26.2_{13.8}$	$37.1_{13.8}$	$46.7_{11.5}$	$66.4_{9.0}$
Data1-100%	$34.6_{15.3}$	$36.4_{17.9}$	$47.4_{17.0}$	$65.6_{10.3}$
Data2-13%	$11.5_{24.4}$	$25.6_{23.9}$	$40.0_{27.0}$	$21.5_{25.6}$
Data2-25%	$36.9_{29.7}$	$35.8_{23.4}$	$40.5_{26.7}$	$46.7_{25.6}$
Data2-50%	$15.0_{14.5}$	$34.4_{24.2}$	$38.4_{17.0}$	$49.8_{26.4}$
Data2-100%	$16.5_{16.2}$	$33.4_{16.4}$	$38.9_{15.9}$	$45.7_{28.1}$

4.1   Results and Discussion

Semi-supervised Pathology Segmentation: We evaluate the APD-Net performance in a semi-supervised experiment by altering the pathology supervision percentage, as seen in Table 1 respectively for the two datasets. For clarity we omit anatomy segmentation results, which are approximately 78% for both Cascaded U-Net and the proposed APD-Net for *Data1* and 64% for *Data2*.

Infarct segmentation is a challenging task, and thus all results of *Data1* and *Data2* present relatively high standard deviation, in agreement with previous literature [7]. In *Data1*, APD-Net consistently improves the Dice score of infarct prediction for all amounts of supervision, compared with both the *Cascaded U-Net* and the *U-Net (unmasked)*. Furthermore, the performance of APD-Net on small amounts of pathology labels is equivalent to the fully supervised setting.

Segmenting pathology in *Data2* is harder, as evidenced by the lower mean and higher standard deviation obtained from all methods. This could be due to the supervised methods overfitting to the smaller dataset size. APD-Net, however, overcomes this issue with semi-supervised training, and outperforms the *U-Net (masked)* at 13% annotations. While at 100% annotations, APD-Net achieves the equivalent Dice as 13% but reduces the standard deviation. More importantly, APD-Net outperforms the *Cascaded U-Net* in all setups demonstrating the benefit of image reconstruction (see also ablation studies later). Examples of correct and unsuccessful segmentations from APD-Net can be seen in Fig. 3, where the existence of sparsely-distributed annotations (right panel of Fig. 3) negatively affects supervised training.

Ablation Studies: We evaluate the effects of critical components including the pathology-masked image reconstruction, disentanglement, teacher-forcing, and the ratio-based triplet loss with 13% and 100% infarct annotations on *Data1*. To evaluate disentanglement, we remove the modality encoder, and allow the anatomy factor to encode continuous values, $s^i \in [0,1]^{H \times W \times C}$. The results presented in Table 2 show that canceling any of the ablated components hurts

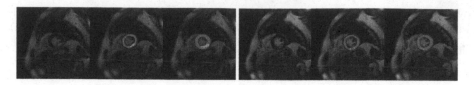

Fig. 3. Segmentation examples from *Data2*. The two panels show a good and failure infarct segmentation case. For each sample, the left image shows the input, and the next two overlay real and predicted myocardium and infarct respectively.

segmentation (except for the masked reconstruction, which at 100% performs as the proposed APD-Net). In particular, reducing annotations at 13% further decreases performance of the ablated models.

Table 2. Ablation studies on *Data1* on two infarct annotation levels (% infarct): no mask reconstruction ($\lambda_M = 0$); no disentanglement (w.o. Disent.); no teacher-forcing strategy ($\lambda^{RA} = \lambda^{RP} = 0$); and no ratio-based triplet loss ($\lambda_{RT} = 0$).

% annotations	$\lambda_M = 0$	w.o. Disent.	$\lambda^{RA} = \lambda^{RP} = 0$	$\lambda_{RT} = 0$	APD-Net (proposed)
13%	$41.4_{15.2}$	$14.9_{8.3}$	$40.7_{12.4}$	$38.8_{14.9}$	$45.3_{14.4}$
100%	$47.7_{15.9}$	$18.4_{16.6}$	$44.5_{12.9}$	$40.3_{10.7}$	$47.4_{17.0}$

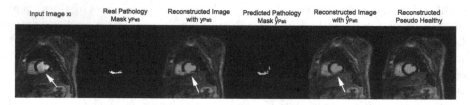

Fig. 4. Reconstruction visualizations with real, predicted pathology masks, and pseudo-healthy. White arrows point at infarct regions.

Figure 4 depicts the effects the disentangled pathology factor on the reconstructed image. Arrows indicate the infarct region, evident when using either the real or predicted pathology to reconstruct the image. In contrast, the infarct is missing when the pathology factor is nulled p_0^i, producing a pseudo-healthy image. Qualitatively, the synthetic image of the proposed APD-Net is similar to the one presented in [3]. The difference between the reconstructed and pseudo-healthy images is driven by the pathology factor. It is enhanced by the ratio based triplet loss that is essential for the desired pathology disentanglement.

5 Conclusions and Future Work

In this paper, we proposed the Anatomy Pathology Disentanglement Network (APD-Net) disentangling the latent space into anatomy, modality, and pathology factors. Trained in a semi-supervised scenario with reconstruction losses enabled by disentangled representation learning and joint optimization losses, APD-Net is capable of segmenting the pathology region effectively when partial pathology annotations are available. APD-Net has shown promising results in pathology segmentation using partial annotations and improved performance compared to other related baselines in the literature.

However, APD-Net still follows the cascaded pathology segmentation strategy that can propagate errors from the first to the second segmentation, which is not solved fundamentally even with the *Teacher-Forcing*. Diseases that deform the anatomical structure (e.g., brain tumour) cannot be predicted by cascading. In addition, we only tested the proposed APD-Net in myocardial infarct, where the pathology manifests as high-intensity regions within the myocardium. As future work, we aim to explore direct pathology segmentation methods without predicting the relevant anatomy. Meanwhile, we plan to investigate extensions of the current APD-Net that are more general and do not restrict to myocardial infarct, while also engaging multi-modal images that would offer complementary anatomical information. Finally, we will test our method on public datasets with more examples to further validate pathology segmentation performance.

Acknowledgement. This work was supported by US National Institutes of Health (1R01HL136578-01). This work used resources provided by the Edinburgh Compute and Data Facility (http://www.ecdf.ed.ac.uk/). S.A. Tsaftaris acknowledges the Royal Academy of Engineering and the Research Chairs and Senior Research Fellowships scheme.

References

1. Abraham, N., Khan, N.M.: A novel focal Tversky loss function with improved attention U-Net for lesion segmentation. In: 2019 IEEE 16th International Symposium on Biomedical Imaging. (ISBI 2019), pp. 683–687. IEEE (2019)
2. Chartsias, A., et al.: Factorised spatial representation learning: application in semi-supervised Myocardial segmentation. In: Frangi, A.F., Schnabel, J.A., Davatzikos, C., Alberola-López, C., Fichtinger, G. (eds.) MICCAI 2018. LNCS, vol. 11071, pp. 490–498. Springer, Cham (2018). https://doi.org/10.1007/978-3-030-00934-2_55
3. Chartsias, A., et al.: Disentangled representation learning in cardiac image analysis. Med. Image Anal. **58**, 101535 (2019)
4. Chartsias, A., et al.: Disentangle, align and fuse for multimodal and zero-shot image segmentation. arXiv preprint arXiv:1911.04417 (2019)
5. Dey, R., Hong, Y.: CompNet: complementary segmentation network for brain MRI extraction. In: Frangi, A.F., Schnabel, J.A., Davatzikos, C., Alberola-López, C., Fichtinger, G. (eds.) MICCAI 2018. LNCS, vol. 11072, pp. 628–636. Springer, Cham (2018). https://doi.org/10.1007/978-3-030-00931-1_72
6. Huang, X., Liu, M.Y., Belongie, S., Kautz, J.: Multimodal unsupervised image-to-image translation. In: Proceedings of the European Conference on Computer Vision (ECCV), pp. 172–189 (2018)
7. Karim, R., et al.: Evaluation of state-of-the-art segmentation algorithms for left ventricle infarct from late Gadolinium enhancement MR images. Med. Image Anal. **30**, 95–107 (2016)

8. Lin, T.Y., Goyal, P., Girshick, R., He, K., Dollár, P.: Focal loss for dense object detection. In: Proceedings of the IEEE international conference on computer vision, pp. 2980–2988 (2017)
9. Moccia, S., et al.: Development and testing of a deep learning-based strategy for scar segmentation on CMR-LGE images. Magn. Reson. Mater. Phys. Biol. Med. **32**(2), 187–195 (2018). https://doi.org/10.1007/s10334-018-0718-4
10. Morshid, A., et al.: A machine learning model to predict hepatocellular carcinoma response to transcatheter arterial chemoembolization. Radiol. Artif. Intell. **1**(5), e180021 (2019)
11. Pang, Y., Hu, D., Sun, M.: A modified scheme for liver tumor segmentation based on cascaded FCNs. In: Proceedings of the International Conference on Artificial Intelligence, Information Processing and Cloud Computing, pp. 1–6 (2019)
12. Qin, C., Shi, B., Liao, R., Mansi, T., Rueckert, D., Kamen, A.: Unsupervised deformable registration for multi-modal images via disentangled representations. In: Chung, A.C.S., Gee, J.C., Yushkevich, P.A., Bao, S. (eds.) IPMI 2019. LNCS, vol. 11492, pp. 249–261. Springer, Cham (2019). https://doi.org/10.1007/978-3-030-20351-1_19
13. Ronneberger, O., Fischer, P., Brox, T.: U-Net: convolutional networks for biomedical image segmentation. In: Navab, N., Hornegger, J., Wells, W.M., Frangi, A.F. (eds.) MICCAI 2015. LNCS, vol. 9351, pp. 234–241. Springer, Cham (2015). https://doi.org/10.1007/978-3-319-24574-4_28
14. Roth, K., Konopczyński, T., Hesser, J.: Liver lesion segmentation with slice-wise 2D Tiramisu and Tversky loss function. arXiv preprint arXiv:1905.03639 (2019)
15. Salehi, S.S.M., Erdogmus, D., Gholipour, A.: Tversky loss function for image segmentation using 3D fully convolutional deep networks. In: Wang, Q., Shi, Y., Suk, H.-I., Suzuki, K. (eds.) MLMI 2017. LNCS, vol. 10541, pp. 379–387. Springer, Cham (2017). https://doi.org/10.1007/978-3-319-67389-9_44
16. Schroff, F., Kalenichenko, D., Philbin, J.: FaceNet: a unified embedding for face recognition and clustering. In: Proceedings of the IEEE Conference on Computer Vision and Pattern Recognition, pp. 815–823 (2015)
17. Tran, G.S., Nghiem, T.P., Nguyen, V.T., Luong, C.M., Burie, J.C.: Improving accuracy of lung nodule classification using deep learning with focal loss. J. Healthc. Eng. **2019** (2019)
18. van Tulder, G., de Bruijne, M.: Learning cross-modality representations from multi-modal images. IEEE Trans. Med. Imaging **38**(2), 638–648 (2018)
19. Williams, R.J., Zipser, D.: A learning algorithm for continually running fully recurrent neural networks. Neural Comput. **1**(2), 270–280 (1989)
20. Xia, T., Chartsias, A., Tsaftaris, S.A.: Pseudo-healthy synthesis with pathology disentanglement and adversarial learning. Med. Image Anal. **64**, 101719 (2020)
21. Zakharov, S., Kehl, W., Planche, B., Hutter, A., Ilic, S.: 3D object instance recognition and pose estimation using triplet loss with dynamic margin. In: 2017 IEEE/RSJ International Conference on Intelligent Robots and Systems (IROS), pp. 552–559. IEEE (2017)

Domain Generalizer: A Few-Shot Meta Learning Framework for Domain Generalization in Medical Imaging

Pulkit Khandelwal[1,2(✉)] and Paul Yushkevich[2]

[1] Department of Bioengineering, University of Pennsylvania, Philadelphia, PA, USA
pulks@seas.upenn.edu
[2] Penn Image Computing and Science Laboratory, Department of Radiology,
University of Pennsylvania, Philadelphia, PA, USA

Abstract. Deep learning models perform best when tested on target (test) data domains whose distribution is similar to the set of source (train) domains. However, model generalization can be hindered when there is significant difference in the underlying statistics between the target and source domains. In this work, we adapt a domain generalization method based on a model-agnostic meta-learning framework [1] to biomedical imaging. The method learns a domain-agnostic feature representation to improve generalization of models to the unseen test distribution. The method can be used for any imaging task, as it does not depend on the underlying model architecture. We validate the approach through a computed tomography (CT) vertebrae segmentation task across healthy and pathological cases on three datasets. Next, we employ few-shot learning, i.e. training the generalized model using very few examples from the unseen domain, to quickly adapt the model to new unseen data distribution. Our results suggest that the method could help generalize models across different medical centers, image acquisition protocols, anatomies, different regions in a given scan, healthy and diseased populations across varied imaging modalities.

Keywords: Domain adaptation · Domain generalization · Meta learning · Vertebrae segmentation · Computed tomography

1 Introduction and Background

In biomedical imaging, deep learning models trained on one dataset are often hard to generalize to other related datasets. Generally, biomedical images can be represented as points on a high-dimensional non-linear manifold. Failure of segmentation and classification algorithms to generalize across imaging modalities,

Electronic supplementary material The online version of this chapter (https://doi.org/10.1007/978-3-030-60548-3_8) contains supplementary material, which is available to authorized users.

© Springer Nature Switzerland AG 2020
S. Albarqouni et al. (Eds.): DART 2020/DCL 2020, LNCS 12444, pp. 73–84, 2020.
https://doi.org/10.1007/978-3-030-60548-3_8

patients, image acquisition protocols, medical centers, healthy and diseased populations, age, etc., can be explained by significant differences in the statistical distributions of datasets on the image manifolds, known as *covariate shift* [2]. Addressing covariate shift by retraining deep learning models on each new data domain is impractical in most applications because of the scarcity of expert labeled data. Therefore, it is important to develop deep learning methods that generalize well to new related datasets not seen during training using few or no annotated examples from the new dataset. *Domain adaptation* [3] and *domain generalization* [4] paradigms aim at reducing the covariate shift between the training and test distributions by learning domain invariant features. Domain adaptation learns a feature representation that is invariant to the statistics of the source and target domains, and is discriminative enough for the actual learning task. Domain adaptation could either be unsupervised or semi-supervised. Domain generalization, a relatively less studied and harder problem, trains models using a variety of source domains to learn a generic feature representation which should perform well on unseen target domains. This flavor of transfer learning does not use any samples from the target distribution during training. Relatedly, few-shot learning is a paradigm which adapts a trained model to a completely new data distribution with very limited labeled training examples [5].

The biomedical imaging community has witnessed several applications of domain adaptation and few-shot learning. Adaptation across different medical centers is a known challenge in image segmentation [6], and has been achieved through both unsupervised [7], and supervised approaches [8]. Cross-modality domain adaptation methods between magnetic resonance (MR) and CT images have been proposed using variational autoencoders [9] for whole heart segmentation, and CycleGANs for segmentation of the prostate [10,11]. Decision forests have been employed to adapt between in-vivo and in-vitro images [12] for intravascular ultrasound tissue segmentation. A few-shot network [13] was proposed to segment multiple organs in MR images. However, a priori knowledge of the unseen test domain is not always available, which hinders model generalizability as discussed above.

Very recently, some groups explored domain generalization for biomedical imaging. (1) A series of nine data augmentation techniques were applied to the training domains to mimic the test distribution [14] for heart ultrasound, heart and prostate MR image segmentation. (2) An episodic training-based meta-learning method [15] was applied to segment brain tissue in T1-weighted MRI across four medical centers. (3) A variational auto-encoder [16] was used to learn three latent subspaces to generalize across patients for a 2D cell segmentation task via domain disentanglement. These methods have certain limitations respectively: (1) In [14], the training data is very large and heavily augmented, which might not be the general case for many problems in medical imaging where the goal is to extract enough information from very limited data for domain generalization; the method is not tested on diseased populations which could vary significantly in anatomies and shapes from a generic healthy population. (2) In [15], again, the training set contains similar anatomies in both the train and test sets; the average performance is quite marginal than the compared baseline, with an improvement of only around 0.8% in Dice score (Baseline: 90.6% vs proposed: 91.4%); does not evaluate on cases with atrophied or irregular brain anatomies. (3) In [16], the method uses domain labels as an additional cue, which we argue is difficult to define precisely, due to its wide range of interpretation; the average performance is not significant with an improvement of 0.4% in average Dice score (Baseline: 95.4% vs proposed: 95.8%); the method evaluates only one 2D dataset, not extending to the 3D case; each patient is considered as different domain acquired at the same medical center, and hence the test and train sets might have similar statistical distributions.

Contributions. In the present work, we extend a gradient-based meta-learning domain generalization method (MLDG) [1] that has shown promise for image classification tasks to the context of biomedical image segmentation, termed **MLDG-Seg**. To evaluate this approach, we focus on the problem of vertebrae segmentation in CT images and utilize three publicly available databases. To address the above-mentioned drawbacks of existing domain generalization methods, we construct three domain generalization contexts: (a) *generalization to new anatomies*: the vertebrae are divided into four domains: lumbar, lower, middle, and upper thoracic regions. The model is trained on three domains, and then tested on the unseen fourth domain. (b) *generalization to a diseased population with fractured vertebrae, dislocated discs*: the model is trained on a healthy population, then tested on unseen data of a diseased population from *another medical center*. (c) *generalization to unseen anatomies, surgical implants, different acquisition protocols, arbitrary orientations, and field of view (FoV)*: the model is tested on a very large dataset. Through these three contexts, we show that MLDG-Seg is able to learn generalized representations from very *limited training examples*. Finally, we show that the learned generalized representation can be quickly adapted with a few examples from the unseen target distribution in a k-shot learning setting to achieve additional performance gains.

2   Methodology

Meta-learning, or *learning to learn* [17], aims at learning a variety of tasks, and then quickly adapting to new tasks in different settings. We adopt the optimization-based model-agnostic method proposed in [1], called Meta Learning Domain Generalization (MLDG). Here, we briefly describe the method.

Description. Let there be two distributions: source S, and target T. Both S, and T share the same task, for example, segmentation or classification with the same label space. The goal of MLDG is to learn a single set of model parameters θ via gradient descent and two meta-learners: meta-train and meta-test procedures. The model is trained on only the source S domains, and then tested on target T domain. The source domains S are split into two sets: meta-train

Algorithm 1. MLDG

1: **Input: Source domains** $\mathcal{S}$
2: Model parameters θ and Hyperparameters: α, β, γ.
3: **for** *iterations* $= 1, 2, \ldots$ **do**
4:     **Randomly Split** source domains $\mathcal{S}$ into meta-train $\hat{S}$, and meta-test $\bar{S}$
5:     **Meta-train**: Gradients $\nabla_\theta = F'_\theta(\hat{S}; \theta)$
6:     Updated parameters: $\theta' \leftarrow \theta - \alpha \nabla_\theta$
7:     **Meta-test**: Compute Loss with updated parameter θ' as $G(\bar{S}; \theta')$
8:     **Final Model parameters**: $\theta \leftarrow \theta - \gamma \frac{\partial(F(\hat{S}; \theta) + \beta G(\bar{S}; \theta - \alpha \nabla_\theta))}{\partial \theta}$
9: **end for**

domains $\hat{S}$, and meta-test domains $\bar{S} = \mathcal{S} - \hat{S}$. The goal of the two splits is to mimic the setting of domain shifts, and thereby make it easier for the model to generalize on an unseen target domain $\mathcal{T}$. We reproduce the learning procedure in Algorithm 1, and show the extended version in Fig. 1.

Explanation. Let's consider a motivating example (carried out later in Experiment 1) from the image manifold of lumbar ($\mathcal{D}1$), lower thoracic ($\mathcal{D}2$) and middle thoracic regions ($\mathcal{D}3$), comprising the set of source domains $\mathcal{S}$; as well as upper ($\mathcal{D}4$) thoracic region, which constitutes the unseen test $\mathcal{T}$ domain. Refer to Algorithm 1. The MLDG method is supposed to learn a single model parameter θ with the help of two optimization steps. At every iteration, the set of images in the source domains, here ($\mathcal{D}1$, $\mathcal{D}2$, and $\mathcal{D}3$) are randomly split into a meta-train (for example, consisting of images from $\mathcal{D}1$, $\mathcal{D}2$ set), and meta-test with the set $\mathcal{D}3$. Now, two losses are computed. The first loss $\mathcal{F}$ is computed using the training examples from meta-train set and the gradient is computed with respect to the model parameter θ. The second loss $\mathcal{G}$ is computed on the meta-test set with the updated parameter $\theta' = \theta - \alpha \nabla_\theta$. The key idea by introducing the second loss in the meta-test stage is that an improvement of the model's performance on the meta-train set should also improve the model's performance on

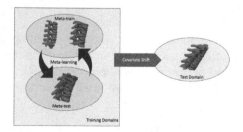

Fig. 1. MLDG-Seg: A schematic. Our method extends the MLDG algorithm to a segmentation by using a 3D Unet-like architecture as the backbone. To explain the segmentation procedure, consider the four domains: lumbar ($\mathcal{D}1$), lower ($\mathcal{D}2$), middle ($\mathcal{D}3$), upper ($\mathcal{D}4$) thoracic regions. As an example, the domains $\mathcal{D}1$, and $\mathcal{D}2$ comprise the meta-train set, and $\mathcal{D}3$ the meta-test. The domain $\mathcal{D}4$ is the held-out test domain.

the meta-test set. The final model parameter θ is updated by taking the gradient of the weighted combination of the two losses $\mathcal{F}$, and $\mathcal{G}$. By doing so, the model is tuned in such a way that performance is improved in both meta-train, and meta-test domains. In other words, the model is regularized and does not overfit to one particular domain, by finding the best possible gradient direction due to the joint optimization of the two losses. Compare this to a "vanilla" setup, where a model is directly given images from the three domains $\mathcal{D}1$, $\mathcal{D}2$, and $\mathcal{D}3$ without any meta-learning setup, might overfit to a domain by minimizing its loss, and maximizing the loss for the other domains.

3   Experiments

3.1   Databases

We validate our approach on three publicly available CT vertebrae segmentation datasets. See supplementary material for sample images of the three datasets.

CSI Challenge - Healthy Cases. The datasets of spine CT (MICCAI 2014 challenge [18]) were acquired during daily clinical routine work in a trauma center from 10 adults (age: 16 to 35 years). In each subject, all 12 thoracic and 5 lumbar vertebrae were manually segmented to obtain ground truth.

xVertSeg Segmentation Challenge - Pathology Cases. This database consists of fractured and non-fractured CT lumbar spine images. We used the 15 subjects, ranging in age from 40 to 90 years, made publicly available with their corresponding lumbar ground truth segmentations [19].

VerSe MICCAI Segmentation Challenge 2020 - Versatile Dataset. This database consists of labeled lumbar, thoracic, cervical vertebrae across 100 cases in the released set [20,21] as of June 2020. The data comes from several medical centers, with a very wide range of acquisition protocols, certain vertebrae with surgical implants, and a range of FoV with arbitrary image orientations.

3.2   Experimental Setup

We define the different implemented procedures, and then perform three experiments. **Baseline** is the vanilla 3D Unet-like [23] architecture. The model is trained on the source domains, and then tested on the unseen held-out test domain. The procedure is then repeated for the other domains. **MLDG-Seg** is the vanilla 3D Unet-like architecture trained under the scheme in Algorithm 1. The model is trained on the source domains, but at every iteration the source domains are divided into meta-train and meta-test splits. The model is then tested on unseen held-out test domain. The procedure is then repeated for the other domains. **k-shot learning** Once the MLDG-Seg model has been trained on source domains, it can be quickly adapted via fine-tuning using very limited labeled examples from the unseen target domain. The weights of the encoder, and bottleneck layers of the 3D Unet-like architecture were frozen in the k-shot learning experiments for fine-tuning. Our approach of k-shot learning is different than that of *test time adaptation* methods [24–27], where usually the encoder weights are not frozen in a Y-shaped network architecture, but are explicitly updated using a few examples at the time of testing. In contrast, we fine-tune the decoder weights with k examples from the test distribution. **Oracle** To establish a theoretical upper bound on segmentation accuracy, we train the vanilla 3D Unet-like architecture using labeled examples from the target domain, and then test on held-out examples from the target domain. This is presumably the easiest task, since the training and test domain distribution are similar. However, in our relatively small datasets the amount of available data for the oracle experiment is smaller than for the domain generalization experiments, so oracle performance reported below should be read with caution.

Experiment 1. The images in all the subjects of the CSI database are divided into different regions: lumbar (L1–L5), lower thoracic (T9–T12), middle thoracic (T5–T8), and upper thoracic vertebrae (T1–T4). Each of these regions comprise of a **domain**. Here, the aim is to let the model generalize across the underlying vertebral anatomy, the intensity profile and the surrounding intervertebral disc space which varies significantly along the vertebral column.

Experiment 2. Here, the four domains i.e., lumbar, lower, middle, and upper thoracic vertebrae from the CSI healthy database comprise the set of source domains. The xVertSeg database with pathology cases comprise the unseen target domain. Here, the aim is to let the model generalize from the vertebrae of healthy subjects to the structure of fractured or dislocated vertebrae images obtained at a different medical center.

Experiment 3. Here, the model is trained on CSI and xVertSeg datasets and tested on VerSe dataset, with the aim to generalize to unseen anatomies, different acquisition protocols, arbitrary orientations, FoVs, and pathology.

Train, Validation, and Test Data Split-up Details. The supplementary material details the training, validation, and test sets for all three databases.

Implementation Details. All the images were converted to $1\,mm^3$ isotropic resolution using FreeSurfer [22]. The images were standardized, using mean subtraction and division by standard deviation, and then normalized between 0 and 1. We implemented a 3D Unet-like [23] architecture in PyTorch [28]. This 3D Unet is the backbone architecture used for all the experiments. Each of the three hyperparameters α, β, and γ in MLDG (Algorithm 1) are set to 1. We use stochastic gradient descent as the optimizer with a learning rate of 0.001, momentum of 0.9, and a weight decay of 5×10^{-5}, dropout with probability of 0.3, and *groupnorm* as the normalization technique. See supplementary material for more details on the network architecture. The loss function used is Generalized Dice Loss [29]. We randomly sample 50 patches of size $64 \times 64 \times 64$ from each subject in a domain, and perform data augmentation by randomly rotating and flipping 30 patches out of these 50 patches for every procedure. For the k-shot procedure, we randomly sampled 5 patches from each of the k-th subject from the unseen distribution. We trained every procedure, as described above, in experiment 1 for 10 epochs, and experiments 2 and 3 for 15 epochs. We use the model that gave the highest Dice score on the validation set to evaluate the unseen test domain. The models were trained on Nvidia P100 Tesla GPUs. A sliding window approach was used to obtain the predicted segmentations, which were then post-processed by retaining the largest connected component.

Evaluation details. We compute Dice coefficient (%), a volume-based measurement, and Average Symmetric Surface Distance (ASSD) in mm, a surface-based metric between the groundtruth image and a given model procedure segmentation output. *Desired: Higher Dice, and lower ASSD scores.* Furthermore, we perform pairwise Wilcoxon signed-rank test [30] between the baseline and each of the other procedures in all the three experiments for both Dice score and ASSD. In the three Tables 1, 2 and 3, we highlight the procedures which reach significance at an $\alpha = 0.05$ significance value using the following notation to denote the level of significance: * ($p <0.05$), ** ($p <0.005$), and *** ($p <0.0005$).

4 Results and Discussion

Experiment 1. Table 1 tabulates the results on different held-out test distributions. Each model is trained on three domains and then tested on the fourth unseen domain, except the oracle which is trained and tested on the same domain. MLDG-Seg consistently outperforms the baseline on both Dice and ASSD, and shows the desired low variance amongst the subjects. With a very few labeled examples from the test distribution, we see a further boost in performance. Figure 2C shows that the MLDG-Seg is better able to segment the region of interest (ROI) by having significantly reduced background error than the baseline, in addition to the correct delineation of intervertebral discs (IVDs) when the held-out distribution is the middle thoracic region. A further improvement in performance is obtained using additional k-shot examples over MLDG-Seg. Ideally, the oracle should have the best performance than the rest of the procedures as the test domain distribution is similar to the training domain

Table 1. Dice score (%), and ASSD in mm (mean ± std. dev) for Experiment 1. Each row reports the result on held-out unseen test domain, where the model was trained on the remaining three domains. Supplement contains results on additional subjects.

Test domain	Baseline	MLDG-Seg	$k=1$	$k=2$	Oracle
Lumbar	81.67 ± 8.45 (1.85 ± 0.75)	87.85 ± 2.73 (1.42 ± 0.31)	88.57 ± 1.53 (1.46 ± 0.24)	88.19 ± 2.47 (1.69 ± 0.67)	83.60 ± 2.68 (2.74 ± 0.37)
Lower Th	83.52 ± 4.12 (2.66 ± 0.98)	86.17 ± 2.17 (1.44 ± 0.09)	81.44 ± 2.46 (2.73 ± 0.49)	82.28 ± 1.49 (2.74 ± 0.55)	80.25 ± 4.16 (2.32 ± 0.41)
Middle Th	55.72 ± 4.13 (10.41 ± 0.42)	64.36 ± 11.45 (6.85 ± 3.18)	75.57 ± 5.60 (3.56 ± 1.64)	76.98 ± 8.66 (2.20 ± 1.45)	83.60 ± 0.58 (2.21 ± 0.54)
Upper Th	82.00 ± 1.45 (1.68 ± 0.18)	83.70 ± 2.19 (1.46 ± 0.20)	74.47 ± 6.21 (4.83 ± 2.56)	81.50 ± 4.07 (1.75 ± 0.39)	75.84 ± 4.00 (1.93 ± 0.36)

distribution. Here in this particular experimental setup, it is *not* surprising to see that the MLDG-Seg (and in some test domains, the baseline) performs better than the oracle. This might be due to the limited number of training (and validation) examples available to train the oracle. Furthermore, since all the procedures were trained for a fixed number of epochs, the oracle might have had a disadvantage of being trained for a lesser number of gradient steps than the MLDG-Seg procedure. An alternative approach would be to train a single oracle model which learns from all the domains simultaneously, and then test on each domain separately.

Experiment 2. Table 2 and the Spaghetti plots in Fig. 3 shows that the MLDG-Seg outperforms the baseline. Figure 2B shows that MLDG-Seg is able to delineate the vertebrae by *not* segmenting the undesired spinal cord, as incorrectly segmented by the baseline. The k-shot setting further improves the segmentation. Here again, it is *not* necessarily surprising to see that MLDG-Seg performs better than the oracle. The oracle was trained with a smaller number of subjects than the baseline and MDG-Seg (see supplement).

Table 2. Dice score (%), and ASSD in mm (mean ± std. dev) for Experiment 2. The model was trained on CSI dataset and tested on xVertSeg dataset, thereby generalizing to a pathology dataset when trained on a healthy population.

Baseline	MLDG-Seg	$k=1$	$k=2$	$k=3$	$k=4$	Oracle
74.13 ± 13.69 (3.16 ± 1.05)	75.34 ± 12.10 (2.42 ± 0.70**)	76.88 ± 7.35 (2.74 ± 0.94*)	76.51 ± 8.85 (2.31 ± 0.65**)	77.95 ± 8.79* (2.38 ± 0.74**)	79.20 ± 8.01* (2.29 ± 0.77**)	74.94 ± 7.84 (3.98 ± 1.62)

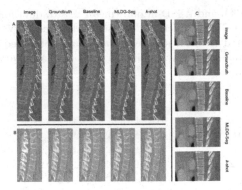

Fig. 2. Qualitative illustrations for the three experiments. (A) Experiment 3: tested on VerSe dataset, here: $k = 7$. (B) Experiment 2: tested on xVert dataset, here: $k = 4$. (C) Experiment 1: shown is the middle thoracic region as the test set, here: $k = 2$. Qualitative results for lumbar, lower, and upper thoracic regions can be found in the supplement. A minor discrepancy at the boundaries is noticed in groundtruth in C, perhaps due to registration errors.

Experiment 3. Table 3 shows the improved performance of MLDG-Seg over the baseline, and consistent performance boost in the few-shot learning regime. The k-shot procedure is either on par with the oracle or outperforms the same for $k = 4, 5$ and 6. Figure 2A depicts the superior performance of MLDG-Seg over the baseline on a compression fraction subject from a different distribution than the training set, where MLDG-Seg is able to segment more vertebrae completely than the baseline. The k-shot setting further improves the performance by segmenting the remaining vertebrae not segmented by baseline or MLDG-Seg. Here, the oracle performs better than most of the procedures, which is due to the fact that the training set for oracle is relatively larger than in experiments 1 and 2.

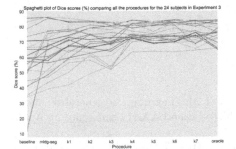
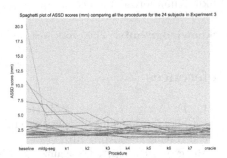

Fig. 3. Spaghetti plots are shown for Experiment 3. *Left*: Dice coefficient, *Right*: ASSD score. Y-axis shows the Dice (%), and the ASSD (mm) scores for the different procedures shown on the X-axis. Each of the plotted lines denote a subject. Therefore, one can track the performance of a procedure for a given subject. See supplementary material for the spaghetti plots for all the other experiments.

Table 3. Dice score (%), and ASSD in mm (mean ± std. dev) for Experiment 3. The model was trained on CSI, and xVertSeg dataset and tested on VerSe dataset, thereby generalizing to fractured or dislocated vertebrae images obtained at different medical centers, unseen anatomies, different acquisition protocols, arbitrary orientations, FoVs, and various pathology.

Procedure	Dice score	ASSD
Baseline	54.58 ± 18.16	4.23 ± 4.21
MLDG-Seg	64.82 ± 13.13***	2.99 ± 1.59*
$k=1$	69.12 ± 10.48***	2.57 ± 1.21**
$k=2$	71.49 ± 9.04***	2.31 ± 0.91***
$k=3$	70.18 ± 9.53***	2.42 ± 0.76**
$k=4$	75.90 ± 4.99***	2.29 ± 0.69*
$k=5$	75.27 ± 5.11***	2.31 ± 0.68*
$k=6$	75.54 ± 5.71***	2.13 ± 0.60**
$k=7$	75.28 ± 5.67***	2.17 ± 0.57**
Oracle	74.97 ± 5.81***	2.31 ± 0.62*

Conclusion In the present work, we benchmarked the performance of a gradient-based meta-learning domain generalization segmentation method in the context of biomedical image analysis, across a variety of training and test settings. The method was not only able to generalize across multiple medical sites and scanners, which is the most widely studied problem of generalization, but was also able to generalize to newly introduced settings of unseen complex vertebrae anatomies, surrounding inter-vertebral discs space, varying bone and soft tissue intensities distribution, diseased populations, different acquisition protocols, arbitrary orientations, and FoVs, thus resembling actual clinical settings. In future, we will evaluate the method on other modalities such as MRI, and US and compare with the recently proposed domain generalization methods. Our source code, scripts, and dataset-split files are available at: https://github.com/Pulkit-Khandelwal/medical-mldg-seg.

Acknowledgments. This work was supported by NIH grant R01 EB017255.

References

1. Li, D., Yang, Y., Song, Y.Z. Hospedales, T.M.: Learning to generalize: meta-learning for domain generalization. In: Thirty-Second AAAI Conference on Artificial Intelligence, April 2018
2. Meng, Q., Rueckert, D., Kainz, B.: Learning cross-domain generalizable features by representation disentanglement. arXiv preprint arXiv:2003.00321 (2020)
3. Ganin, Y., et al.: Domain-adversarial training of neural networks. J. Mach. Learn. Res. **17**(1), 1–35 (2016)
4. Li, D., Yang, Y., Song, Y.Z. Hospedales, T.M.: Deeper, broader and artier domain generalization. In: Proceedings of the IEEE International Conference on Computer Vision, pp. 5542–5550 (2017)

5. Snell, J., Swersky, K., Zemel, R.: Prototypical networks for few-shot learning. In: Advances in Neural Information Processing Systems, pp. 4077–4087 (2017)
6. Glocker, B., Robinson, R., Castro, D.C., Dou, Q., Konukoglu, E.: Machine learning with multi-site imaging data: an empirical study on the impact of scanner effects. arXiv preprint arXiv:1910.04597 (2019)
7. Kamnitsas, K., et al.: Unsupervised domain adaptation in brain lesion segmentation with adversarial networks. In: Niethammer, M., et al. (eds.) IPMI 2017. LNCS, vol. 10265, pp. 597–609. Springer, Cham (2017). https://doi.org/10.1007/978-3-319-59050-9_47
8. Valindria, V.V., et al.: Domain adaptation for MRI organ segmentation using reverse classification accuracy. arXiv preprint arXiv:1806.00363 (2018)
9. Ouyang, C., Kamnitsas, K., Biffi, C., Duan, J., Rueckert, D.: Data efficient unsupervised domain adaptation for cross-modality image segmentation. In: Shen, D., et al. (eds.) MICCAI 2019. LNCS, vol. 11765, pp. 669–677. Springer, Cham (2019). https://doi.org/10.1007/978-3-030-32245-8_74
10. Liu, Y., et al.: Cross-modality knowledge transfer for prostate segmentation from CT scans. In: Wang, Q., et al. (eds.) DART/MIL3ID -2019. LNCS, vol. 11795, pp. 63–71. Springer, Cham (2019). https://doi.org/10.1007/978-3-030-33391-1_8
11. Dou, Q., et al.: PnP-AdaNet: plug-and-play adversarial domain adaptation network at unpaired cross-modality cardiac segmentation. IEEE Access 7, 99065–99076 (2019)
12. Conjeti, S., et al.: Supervised domain adaptation of decision forests: transfer of models trained in vitro for in vivo intravascular ultrasound tissue characterization. Med. Image Anal. 32, 1–17 (2016)
13. Roy, A.G., Siddiqui, S., Pölsterl, S., Navab, N., Wachinger, C.: 'Squeeze and excite'guided few-shot segmentation of volumetric images. Med. Image Anal. 59, 101587 (2020)
14. Zhang, L., et al.: Generalizing deep learning for medical image segmentation to unseen domains via deep stacked transformation. IEEE Trans. Med. Imaging 39(7), 2531–2540 (2020)
15. Dou, Q., de Castro, D.C., Kamnitsas, K., Glocker, B.: Domain generalization via model-agnostic learning of semantic features. In: Advances in Neural Information Processing Systems, pp. 6450–6461 (2019)
16. Ilse, M., Tomczak, J.M., Louizos, C., Welling, M.: DIVA: Domain invariant variational autoencoders. arXiv preprint arXiv:1905.10427 (2019)
17. Finn, C., Abbeel, P., Levine, S.: Model-agnostic meta-learning for fast adaptation of deep networks. In: Proceedings of the 34th International Conference on Machine Learning, vol. 70, pp. 1126–1135. JMLR. org, August 2017
18. Yao, J., et al.: A multi-center milestone study of clinical vertebral CT segmentation. Comput. Med. Imaging Graph. 49, 16–28 (2016)
19. Korez, R., Ibragimov, B., Likar, B., Pernuš, F., Vrtovec, T.: A framework for automated spine and vertebrae interpolation-based detection and model-based segmentation. IEEE Trans. Med. Imaging 34(8), 1649–1662 (2015)
20. Sekuboyina, A., et al.: VerSe: a vertebrae labelling and segmentation benchmark. arXiv preprint arXiv:2001.09193 (2020)
21. Löffler, M.T., et al.: A vertebral segmentation dataset with fracture grading. Radiol. Artif. Intell. 2(4), e190138 (2020)
22. Fischl, D.: FreeSurfer. Neuroimage 62(2), 774–781 (2012)
23. Ronneberger, O., Fischer, P., Brox, T.: U-Net: convolutional networks for biomedical image segmentation. In: Navab, N., Hornegger, J., Wells, W.M., Frangi, A.F.

(eds.) MICCAI 2015. LNCS, vol. 9351, pp. 234–241. Springer, Cham (2015). https://doi.org/10.1007/978-3-319-24574-4_28

24. Wang, G., et al.: Interactive medical image segmentation using deep learning with image-specific fine tuning. IEEE Trans. Med. Imaging **37**(7), 1562–1573 (2018)

25. Sun, Y., Wang, X., Liu, Z., Miller, J., Efros, A.A., Hardt, M.: Test-time training for out-of-distribution generalization. arXiv preprint arXiv:1909.13231 (2019)

26. Karani, N., Chaitanya, K., Konukoglu, E.: Test-time adaptable neural networks for robust medical image segmentation. arXiv preprint arXiv:2004.04668 (2020)

27. Zhang, J., et al.: Fidelity imposed network edit (FINE) for solving ill-posed image reconstruction. NeuroImage **211**, 116579 (2020)

28. Paszke, A., et al.: PyTorch: an imperative style, high-performance deep learning library. In: Advances in Neural Information Processing Systems, pp. 8024–8035 (2019)

29. Crum, W.R., Camara, O., Hill, D.L.: Generalized overlap measures for evaluation and validation in medical image analysis. IEEE Trans. Med. Imaging **25**(11), 1451–1461 (2006)

30. Wilcoxon, F.: Individual comparisons by ranking methods. In: Breakthroughs in statistics, pp. 196–202. Springer, New York (1992). https://doi.org/10.1007/978-1-4612-4380-9_16

Parts2Whole: Self-supervised Contrastive Learning via Reconstruction

Ruibin Feng[1]([✉]), Zongwei Zhou[1], Michael B. Gotway[2], and Jianming Liang[1]([✉])

[1] Arizona State University, Tempe, AZ 85281, USA
{rfeng12,zongweiz,jianming.liang}@asu.edu
[2] Mayo Clinic, Scottsdale, AZ 85259, USA
Gotway.Michael@mayo.edu

Abstract. Contrastive representation learning is the state of the art in computer vision, but requires huge mini-batch sizes, special network design, or memory banks, making it unappealing for 3D medical imaging, while in 3D medical imaging, reconstruction-based self-supervised learning reaches a new height in performance, but lacks mechanisms to learn contrastive representation; therefore, this paper proposes a new framework for self-supervised contrastive learning via reconstruction, called Parts2Whole, because it exploits the *universal* and *intrinsic* part-whole relationship to learn contrastive representation without using contrastive loss: Reconstructing an image (whole) from its own parts compels the model to learn similar latent features for all its own parts in the latent space, while reconstructing different images (wholes) from their respective parts forces the model to simultaneously push those parts belonging to different wholes farther apart from each other in the latent space; thereby the trained model is capable of distinguishing images. We have evaluated our Parts2Whole on five distinct imaging tasks covering both classification and segmentation, and compared it with four competing publicly available 3D pretrained models, showing that Parts2Whole significantly outperforms in two out of five tasks while achieves competitive performance on the rest three. This superior performance is attributable to the contrastive representations learned with Parts2Whole. Codes and pretrained models are available at github.com/JLiangLab/Parts2Whole.

Keywords: 3D Self-supervised Learning · Contrastive representation learning · Transfer learning

1 Introduction and Related Work

Contrastive representation learning has made a leap in computer vision. For example, MoCo [13] introduces the momentum mechanism, and SimCLR [10] proposes a simple framework for contrastive learning; both methods achieve state-of-the-art results and even outperform supervised ImageNet pretraining. However, contrastive learning requires huge mini-batch sizes [10,14], special network design [3], or memory banks [13,14,19] to store feature representations of

© Springer Nature Switzerland AG 2020
S. Albarqouni et al. (Eds.): DART 2020/DCL 2020, LNCS 12444, pp. 85–95, 2020.
https://doi.org/10.1007/978-3-030-60548-3_9

all images in the dataset, making it unattractive for 3D medical imaging applications. Taking the mini-batch size as an example, SimCLR [10] recommends 8192, which is impractical for 3D image data due to the current GPU memory limitation. On the other hand, reconstruction-based self-supervised learning has proven to be effective and efficient for 3D medical image analysis. Models Genesis [20] establish autodidactic models by restoring images that underwent four transformations. Later, Tao et al. [18] permute volumetric data via 3D voxel rotation and then restore the original data to learn robust features. Therefore, in this paper, we seek to answer the following critical question: *Can we learn contrastive representations via reconstruction for 3D medical imaging to effectively address the aforementioned barriers associated with contrastive learning?*

To answer this question, we exploit a *universal* and *intrinsic* property, the part-whole relationship, where an entire image is regarded as the whole and any of its patches are considered as its parts. This property has been explored in Sim-CLR [10] via contrastive prediction between the global view (whole) and local view (part). Later, SwAV [7] observed that mapping local views to global views can significantly increase the representation quality. However, instead of directly comparing features or their cluster assignments, we reconstruct a whole from its parts with a pair of encoder and decoder. By doing so, the deep model is compelled to learn contrastive representations embedded with part-whole semantics: (1) the representations of parts belonging to the same whole are close, and (2) the representations of parts belonging to different wholes are far away. We refer to our self-supervised learning framework as Parts2Whole.

Notably, Parts2Whole integrates advantages of several existing self-supervised learning approaches, but overcomes their limitations: Parts2Whole (1) discriminates individual images as Exemplar [11] aims for,[1] but overcomes the scalability issue stemmed from classification because of our use of reconstruction; (2) learns contrastive representations like contrastive learning methods [10,13], but eliminate the need for huge mini-batch size or memory bank by avoiding direct feature comparison; (3) exploits recurrent anatomical structures like Models Genesis [20], but enriches feature representations with part-whole semantics. Furthermore, Models Genesis [20], particularly in-painting/out-painting, only interpolates/extrapolates known masked regions since there are pixel-coordinate mappings between inputs and ground truths. However, in Parts2Whole, the restored area is unknown and the pixel-coordinate mapping does not exist because inputs are randomly cropped and resized from ground truths. In other words, Parts2Whole needs to learn the scale of wholes as well as recover the missing contents, which is much harder and therefore yields more powerful source models.

Our pretrained model is extensively evaluated on five distinct medical target tasks and compare with four competing publicly available 3D models pretrained in either fully supervised or self-supervised fashion in Sect. 3.2. The statistical

---

[1] If we consider each whole image itself as a "label", the training process of Parts2Whole is equivalent to predicting the correct "label" given a part of one image as input, or discriminating each image from its parts.

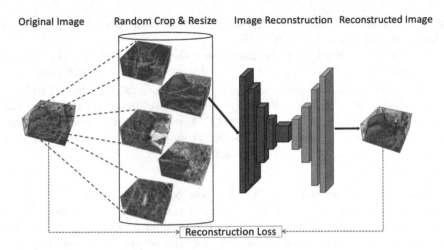

Fig. 1. We propose a new self-supervised learning framework, called Parts2Whole, because it exploits a *universal* and *intrinsic* property, the part-whole relationship, where an entire image is regarded as the whole and its cropped patches are considered as its parts. Parts2Whole aims to learn contrastive representations that embed the part-whole semantics by reconstructing a whole (blue framed) from its resized randomly cropped parts (red framed). To avoid trivial solutions, we crop each whole with random scales and aspect ratios to erase low-level cues across different parts while maintaining informative structures and textures. Additionally, we do not use the skip connections to avoid low-level details passing from the encoder to decoder, yielding generic pretrained models with strong transferability. The model is trained in an end-to-end fashion and the reconstruction loss is measured with Euclidean distance. (Color figure online)

analysis in Table 1 shows Parts2Whole significantly outperforms in two out of five tasks while achieves competitive performance on the rest three. We also empirically validate that Parts2Whole can learn contrastive representations in an image reconstruction framework in Fig. 2 and Sect. 3.3. Finally, our Parts2Whole design is justified by ablating its main components in Table 2.

2 Proposed Method

Our goal is to learn contrastive representations embedded with part-whole semantics by reconstructing the whole image from its parts, with 3D unlabeled images. The proposed self-supervised learning framework is illustrated in Fig. 1.

Problem Formulation: Denote a set of 3D unlabeled images as $\{x_i \in X : i \in [1, N]\}$ where N is the number of whole images. Each image x_i is randomly cropped and resized to generate various parts, referred to as $\{p_i^j \in P_i : i \in [1, N], j \in [1, M)\}$. The task is to predict the whole image x_i from its local patch p_i^j by training a pair of encoder $(\mathcal{F}_E)$ and decoder $(\mathcal{F}_D)$ to minimize the loss function, denoted by $\mathcal{L} = \sum_i \sum_j l(\mathcal{F}_D(\mathcal{F}_E(p_i^j)), x_i)$, where $l(\cdot)$ is a metric measuring the difference between the model outputs and ground truths. We use Euclidean distance as $l(\cdot)$ in this work.

Since the output images are generated via a *shared* decoder $(\mathcal{F}_D)$, the encoder $(\mathcal{F}_E)$ is forced to learn contrastive representations that embed the part-whole semantics. To be specific, after training, $\mathcal{F}_E(p_i^j)$ and $\mathcal{F}_E(p_{i'}^{j'})$ are forced to be as close as possible if $i = i'$ since the two representations are mapped to the same groud truth (x_i) via the shared decoder $(\mathcal{F}_D)$, while far away from each other otherwise since they are mapped to different ground truths. To avoid ambiguous cases, we assume that *no part is also a whole*.

Removing Skip Connection: The skip connection (or the shortcut connection) was proposed in [6] and adopt to connect the encoder and decoder in the U-Net architecture [15]. The goal is to let the decoder access the low-level features produced by the encoder layers such that the boundaries in segmentation maps produced by the decoder could be accurate. However, we argue that if the network can solve the proxy task using lower-level cues, it does not need to learn semantically meaningful representations. Therefore, in proxy task training, we adopt the 3D U-Net architecture[2] and remove the skip connections to force the bottleneck representations encoding high-level information, which is different from [18,20] in the perspective of network architectures. Table 2 demonstrates the effects of skip connections in proxy task training. Notice that although we cannot provide a pretrained decoder as [20] does, our model offers very competitive performance on three segmentation tasks with a randomly initialized decoder, suggesting our pretrained encoder learns strong, generic features.

Extracting Local Yet Informative Parts: The part size is a critical component in our proxy task design. For example, when the crop scale is too large, the task is downgraded to training an autoencoder without learning semantics. On the other hand, the task can be unsolvable if the parts are too small and do not contain enough information. To avoid such degenerate solutions, we restrict the cropped patches covering less than 1/4 area of the whole image. By doing so, the low-level cues across different parts are largely erased. Additionally, we set each part covering more than 1/16 area of the original image to have discriminative structures and textures. We analyze the effects of crop scales in Table 2.

3   Experiments

3.1   Experiment Settings

Proxy Task Training: We pretrain our model on LUNA-2016 [16] dataset without using any label shipped with it. To avoid test data leakage, we use 623 CT scans instead of all 888 scans. We first cropped the original CT scans to small, *non-overlapped* 28,144 sub-scans with dimensions equal to $128 \times 128 \times 64$.

---

[2] 3D U-Net: github.com/ellisdg/3DUnetCNN.

Table 1. Our pretrained model achieves significantly better or at least comparable performance on five distinct medical target tasks over four publicly available 3D models pretrained in both supervised and self-supervised fashion. Each experiment is conducted for 10 trials and summarized with the mean and standard deviation (mean±s.t.d.). The paired t-test results between our method and the previous top-1 solution are tabulated in terms of the p-value. The best approaches are **bolded** while the others are highlighted in blue if they achieve equivalent performance compared with the best one (*i.e.*, $p > 0.05$).

Approach	NCC AUC(%)	NCS IoU(%)	LCS[††] IoU(%)	ECC AUC(%)	BMS[†††] IoU(%)
Scratch	94.25±5.07	74.05±1.97	77.82±3.87	79.99±8.06	63.91±1.41
I3D [8]	98.26±0.27	71.58±0.55	70.65±4.26	80.55±1.11	67.83±0.75
NiftyNet [12]	94.14±4.57	52.98±2.05	83.23±1.05	77.33±8.05	60.78±1.60
MedicalNet [9]	95.80±0.49	75.68±0.32	85.52±0.58	86.43±1.44	66.09±1.35
Models Genesis [20]	97.90±0.57	**77.62±0.64**	84.17±1.93	**87.20±2.87**	68.08±1.15
Parts2Whole (ours)	**98.67±0.23**	77.35±0.61	**86.70±0.62**	86.14±2.97	**68.33±0.41**
p-value[†]	0.0011	0.1709	0.0002	0.2126	0.2654

[†] p-values are calculated between Parts2Whole and the previous top-1 solution.
[††] The IoU score is calculated using binarized masks with a threshold equal to 0.5 to better presented the segmentation quality, while [20] uses the original masks without thresholding.
[†††] Notice the results are different from those reported in [20] since we use *real* data while Models Genesis were evaluated with *synthetic* data.

We treat each generated sub-scan as a *whole* and crop parts from it on the fly. The cropped parts contain [1/16, 1/4] volume of the whole image.

Target Task Training: To extensively evaluate our pretrained 3D model, we follow the practice employed in [20] and investigate five distinct medical applications, including lung nodule false positive reduction (NCC) [16], lung nodule segmentation (NCS) [2], liver segmentation (LCS) [5], pulmonary embolism false positive reduction (ECC) [17], and brain tumor segmentation (BMS) [4].

3.2 Parts2Whole Yields Competitive 3D Pretrained Models

To extensively evaluate our method, we compare Parts2Whole with four *publicly available 3D models* pretrained in both supervised and self-supervised fashion. To be specific, we test two models supervisely pretrained on 3D medical segmentation tasks: NiftyNet [12] with Dense V-Networks and MedicalNet [9] with 3D-ResNet-101 as the backbone. The former is pretrained with a multi-organ CT segmentation task, and the latter is pretrained with an aggregate dataset (*i.e.*, 3DSeg-8) from eight public medical datasets. We also evaluate I3D [8], which is pretrained with natural videos but has been successfully applied for lung cancer classification [1]. For self-supervised learning, we choose Models Genesis [20], the current state of the art in 3D medical imaging, as our baseline.

The experimental results are summarized in Table 1. First of all, we observe that I3D works well on NCC but performs inferiorly on the other 4 tasks. This suboptimal performance may attribute to the marked difference between natural and medical domains. On the other hand, NiftyNet and MedicalNet, which are fully supervised with medical data, also show relatively poor transferability. We hypothesize that the main reason is the limited amount of annotation for supervising. A piece of evidence is that MedicalNet considerably outperforms NiftyNet by aggregating eight datasets for pretraining. These observations highlight the significance of self-supervised learning in the 3D medical domain, which can close the domain gap and utilize the vast amount of unannotated data.

In contrast with fully supervised pretraining, both self-supervised learning methods (Models Genesis and Parts2Whole) achieve promising results on all five target tasks across organs, diseases, datasets, and modalities. Specifically, for NCC and LCS, Parts2Whole not only has higher AUC/IoU scores and lower standard deviations but also significantly outperforms Models Genesis based on the t-test ($p < 0.05$). On the other hand, Models Genesis achieves better performance by a small margin on NCS and ECC tasks. On the BMS task, which has considerable distance from the proxy dataset (*i.e.,* different disease, organ, and modality), Parts2Whole is still competitive compared to other baselines. Last but not least, since Models Genesis provides both pretrained encoder and decoder, one can expect it to have certain advantages on segmentation tasks (*i.e.,* NCS, LCS, and BMS). Nonetheless, Parts2Whole yields promising results on all segmentation tasks with the same architecture (*i.e.,* 3D U-Net) and a randomly initialized decoder, suggesting the encoder pretrained with Parts2Whole learns features with strong transferability. Next, we will experimentally investigate the properties of feature representations learned in Parts2Whole.

3.3   Parts2Whole Learns Contrastive Representations

To understand the feature representations learned with Parts2Whole, we first visualize the t-SNE embeddings of random, Models Genesis, and Parts2Whole features in Fig. 2(a). Specifically, we randomly select 10 whole images and generate 200 parts for each image with a crop scale [1/16, 1]. Each circle represents one part while each diamond represents a whole image (*i.e.,* crop scale is equal to 1). Different colors and circle sizes denote different images and crop scales.

First of all, unlike the entangled features from random initialization, features pretrained with Models Genesis and Parts2Whole are more separable. However, Models Genesis features are not distinguishable especially when the inputs are cropped with small scales (seeing the red-framed part). On the contrary, Parts2Whole features from the same image are well grouped, while those from different images are highly separable regardless of different crop scales. More importantly, although the network is never trained with large patches or the whole image (*i.e.,* crop scale is equal to [1/4, 1]), it correctly aligns all features from the same image together. This magnificent generalization ability suggests that Parts2Whole learns representations that embed the part-whole semantics.

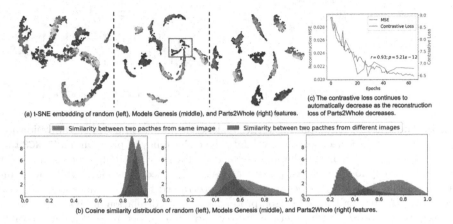

(a) t-SNE embedding of random (left), Models Genesis (middle), and Parts2Whole (right) features.

(c) The contrastive loss continues to automatically decrease as the reconstruction loss of Parts2Whole decreases.

Similarity between two pacthes from same image   Similarity between two pacthes from different images

(b) Cosine similarity distribution of random (left), Models Genesis (middle), and Parts2Whole (right) features.

Fig. 2. To understand the learned representations, we first visualize the t-SNE embeddings of random, Models Genesis, and Parts2Whole features in (a). We use circles to represent parts while diamonds represent whole images. The colors and circle sizes denote the different wholes and crop scales. Compared with random and Models Genesis features, the Parts2Whole features from the same images are well grouped, while features from different images are highly separable, despite the different crop scales. Furthermore, we leverage the entire validation set and measure the cosine similarity between features of two parts belonging to the same or different images in (b). Notice that the similarity distributions of Parts2Whole features are more separable than those of random and Models Genesis features, indicating that Parts2Whole learns better representations. Last but not least, as shown in (c), the contrastive loss continues to automatically decrease validating Parts2Whole can learn contrastive representations.

To further analyze the feature representations, we leverage all the validation images. For each image x_i, we generate 100 pairs of parts from it (referred to as positive pairs). Additionally, we generate 100 negative pairs, while each one contains one part from x_i and one part from another random picked image. We calculate the cosine similarity between each positive pair and negative pair. The similarity distributions of random, Models Genesis, and Parts2Whole features are shown in Fig. 2(b). Note that the similarity distributions of Parts2Whole features are more separable than those of Models Genesis and random features. It further indicates that Parts2Whole learns contrastive representations.

We also investigate the change of the contrastive loss along the training process in Fig. 2(c). We test every whole image 100 times, while in each test, we randomly generate 1 positive pair and 5000 negative pairs for the contrastive loss[3] calculation. As illustrated, the contrastive loss continues to decrease as the reconstruction loss decreases. Additionally, we perform the Pearson product-moment correlation analysis between the reconstruction loss and the contrastive loss.

---

[3] Denote the l_2-normalized features of a positive pair and negative pair as $\{\mathcal{F}_E(p_i), \mathcal{F}_E(p_i')\}$ and $\{\mathcal{F}_E(p_i), \mathcal{F}_E(p_j)\}$, respectively. The contrastive loss is calculated as $-\log \frac{\exp(\mathcal{F}_E(p_i)\cdot\mathcal{F}_E(p_i')/\tau)}{\exp(\mathcal{F}_E(p_i)\cdot\mathcal{F}_E(p_i')/\tau + \sum_{j=1}^{5000}\exp(\mathcal{F}_E(p_i)\cdot\mathcal{F}_E(p_j)/\tau)}$ where $\tau = 0.7$.

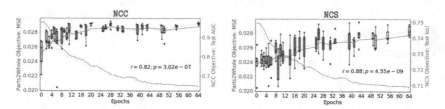

Fig. 3. The overall performance of target tasks continues to improve as the validation loss in the proxy task decreases. We validate the consistency of proxy and NCC/NCS target objective by evaluating 26 checkpoints saved in the proxy training process. It is clear that as the proxy loss decreases, the average AUC/IoU score increases while the standard deviation decreases, suggesting that the pretrained model becomes more generic and robust. Additionally, the Pearson product-moment correlation analysis indicates a strong *positive* co-relationship between proxy and target objectives (Pearson's r-value > 0.5).

The high Pearson's r-values (0.93) suggest a strong positive correlation, validating that Parts2Whole can minimize the contrastive loss and learn contrastive representations with an image reconstruction framework. *Notice that we achieve the goal of contrastive learning with small mini-batch sizes (16 instead of 8192 suggested in [10])*, a general 3D U-Net architecture, and without using memory banks — effectively addressing the barriers associated with previous contrastive learning methods. However, it is still not clear whether good contrastive features embedded with part-whole semantics can yield strong transferability, since the proxy task is agnostic about the target tasks. To answer this question, we systematically investigate the relationship between the reconstruction loss in the proxy task and the test performance in target tasks in the next section.

3.4 Parts2Whole's Objective Is Positively Correlated with Target Objectives

Wu *et al.* [19] suggested that a good proxy task is able to improve the target task performance consistently as the proxy objective is optimized. Following this practice, we validate the consistency of proxy and task objectives by evaluating 26 checkpoints saved in the proxy training process. Specifically, we fine-tune every checkpoint 5 times on NCC and NCS target tasks. To reduce the computational cost, we only use partial training data (45% and 10% for NCC and NCS, respectively). We plot the proxy reconstruction loss and target scores (AUC/IoU) as a function of proxy task training epochs in Fig. 3. We can observe that, as the reconstruction ability in the proxy task improves (*i.e.,* the validation MSE decreases), the transferability of the pretrained model also improves (*i.e.,* the average target score (AUC/IoU) increases while the standard deviation decreases). We further investigate this relationship by performing Pearson product-moment correlation analysis between the proxy objective (*i.e.,* reconstruction quality, measured by (1-MSE)) and target objective (measured by

Table 2. Target task performance on source models pretrained with different proxy task settings. First, removing skip connections (comparing Column 2 to 3) can significantly improve the performance, suggesting skip connections provide shortcuts to solve the proxy task. We also observe that by reducing the cropping scale (from Column 3 to 6), the overall performer continuously increases, plateaus at $[1/16, 1/4]$, and appears to saturate when the scale is less than $1/8$. These observations indicate the importance of crop scales in our proxy task design.

Setting	$[\frac{1}{16}, 1]$ w/ s.c.[†]	$[\frac{1}{16}, 1]$	$[\frac{1}{16}, \frac{1}{2}]$	$[\frac{1}{16}, \frac{1}{4}]$	$[\frac{1}{16}, \frac{1}{8}]$	$[\frac{1}{32}, \frac{1}{16}]$
NCC	88.48±8.24	93.78±2.12	91.48±0.45	**94.84±1.58**	93.52±1.32	91.69±4.12
NCS	70.64±0.21[††]	72.72±0.42	73.29±0.58	**74.23±0.87**	73.43±0.32	73.66±0.36

[†] The source model is trained with skip connections (s.c.) between the encoder and decoder.
[††] We fine-tune both pretrained encoder and decoder on the target task.

AUC/IoU scores). The high Pearson's r-values (0.82 and 0.88 in NCC and NCS, respectively) suggest a strong *positive* co-relationship between proxy and target objectives. This analysis indicates that our superior target performance is attributable to the decreasing of reconstruction loss and the learned contrastive features.

3.5 Ablation Study

A Good Proxy Task Needs to be Hard But Feasible. Our Parts2Whole design contains two main components: removing skip connections and selecting proper crop scales. We ablate the impacts of the two components to justify our proxy task design. We evaluate source models pretrained with different proxy task settings on NCC and NCS target tasks with 45% and 10% training data, respectively. The experimental results are tabulated in Table 2.

First, we study the effects of skip connections in Column 2 to 3 of Table 2. By removing skip connections while keeping the same cropping scale, the target performance improves significantly by 5.30 and 2.08 points in NCC and NCS, respectively. It suggests that skip connections may pass lower-level details from the encoder to decoder, and ergo provide some shortcuts to solve the proxy task.

Second, with the same network architecture (*i.e.*, no skip connections), we study the effects of different part sizes in Column 3–7 of Table 2. When the upper bound of part sizes is gradually reduced, the overall performance continuously increases, plateaus at $1/4$, and appears to saturate at $1/8$. On the other hand, when the parts are too small (*i.e.*, less than $1/16$), the target performance drops by 3.15 and 0.57 points in NCC and NCS, respectively. These observations indicate the importance of proper part sizes in our proxy task design—the parts should be small enough to avoid trivial solutions while large enough to contain enough information to recover the whole images. In other words, we would like to point out that, naturally, *a good proxy task should be hard enough but still feasible.*

4   Conclusion and Future Work

We present a new self-supervised framework, Parts2Whole, by exploiting the *universal* and *intrinsic* part-whole relationship. Our Parts2Whole can learn contrastive representations in an image reconstruction framework. The experimental results show our pretrained model achieves competitive performance over four publicly available pretrained 3D models on five distinct medical target tasks. However, since we only use the part-whole relationship, incorporating other domain knowledge or transformations may boost the results further, as suggested in [10,20]. One promising direction is to include color/intensity transformations since the similar intensity distribution across parts from one image may provide shortcuts to solve the proxy task [10]. On the other hand, Parts2Whole can minimize contrastive loss without explicitly training with it. It points out an intriguing future work—integrating Parts2Whole and contrastive learning into a unified framework to make the leap in the 3D medical imaging domain.

Acknowledgments. This research has been supported partially by ASU and Mayo Clinic through a Seed Grant and an Innovation Grant, and partially by the NIH under Award Number R01HL128785. The content is solely the responsibility of the authors and does not necessarily represent the official views of the NIH. This work has utilized the GPUs provided partially by the ASU Research Computing and partially by the Extreme Science and Engineering Discovery Environment (XSEDE) funded by the National Science Foundation (NSF) under grant number ACI-1548562. We would like to thank Jiaxuan Pang, Md Mahfuzur Rahman Siddiquee, and Zuwei Guo for evaluating I3D, NiftyNet, and MedicalNet, respectively. The content of this paper is covered by patents pending.

References

1. Ardila, D., et al.: End-to-end lung cancer screening with three-dimensional deep learning on low-dose chest computed tomography. Nat. Med. **25**(6), 954–961 (2019)
2. Armato III, S.G., McLennan, G., Bidaut, L., et al.: The lung image database consortium (LIDC) and image database resource initiative (IDRI): a completed reference database of lung nodules on CT scans. Med. Phys. **38**(2), 915–931 (2011)
3. Bachman, P., Hjelm, R.D., Buchwalter, W.: Learning representations by maximizing mutual information across views. In: Advances in Neural Information Processing Systems, pp. 15509–15519 (2019)
4. Bakas, S., Reyes, M., Jakab, A., et al.: Identifying the best machine learning algorithms for brain tumor segmentation, progression assessment, and overall survival prediction in the brats challenge. arXiv preprint arXiv:1811.02629 (2018)
5. Bilic, P., Christ, P.F., Vorontsov, E., Chlebus, G., Chen, H., Dou, Q., et al.: The liver tumor segmentation benchmark (LiTS). arXiv preprint arXiv:1901.04056 (2019)
6. Bishop, C.M., et al.: Neural Networks for Pattern Recognition. Oxford University Press, Oxford (1995)
7. Caron, M., Misra, I., Mairal, J., et al.: Unsupervised learning of visual features by contrasting cluster assignments. arXiv preprint arXiv:2006.09882 (2020)

8. Carreira, J., Zisserman, A.: Quo vadis, action recognition? A new model and the kinetics dataset. In: Proceedings of the IEEE Conference on Computer Vision and Pattern Recognition, pp. 6299–6308 (2017)

9. Chen, S., Ma, K., Zheng, Y.: Med3D: Transfer learning for 3D medical image analysis. arXiv preprint arXiv:1904.00625 (2019)

10. Chen, T., Kornblith, S., Norouzi, M., Hinton, G.: A simple framework for contrastive learning of visual representations. arXiv preprint arXiv:2002.05709 (2020)

11. Dosovitskiy, A., Springenberg, J.T., Riedmiller, M., Brox, T.: Discriminative unsupervised feature learning with convolutional neural networks. In: Advances in Neural Information Processing Systems, pp. 766–774 (2014)

12. Gibson, E., Li, W., Sudre, C., et al.: NiftyNet: a deep-learning platform for medical imaging. Comput. Methods Programs Biomed. **158**, 113–122 (2018)

13. He, K., Fan, H., Wu, Y., Xie, S., Girshick, R.: Momentum contrast for unsupervised visual representation learning. In: Proceedings of the IEEE/CVF Conference on Computer Vision and Pattern Recognition, pp. 9729–9738 (2020)

14. Misra, I., Maaten, L.V.D.: Self-supervised learning of pretext-invariant representations. In: Proceedings of the IEEE/CVF Conference on Computer Vision and Pattern Recognition, pp. 6707–6717 (2020)

15. Ronneberger, O., Fischer, P., Brox, T.: U-Net: convolutional networks for biomedical image segmentation. In: Navab, N., Hornegger, J., Wells, W.M., Frangi, A.F. (eds.) MICCAI 2015. LNCS, vol. 9351, pp. 234–241. Springer, Cham (2015). https://doi.org/10.1007/978-3-319-24574-4_28

16. Setio, A.A.A., Traverso, A., De Bel, T., et al.: Validation, comparison, and combination of algorithms for automatic detection of pulmonary nodules in computed tomography images: the LUNA16 challenge. Med. Image Anal. **42**, 1–13 (2017)

17. Tajbakhsh, N., Gotway, M.B., Liang, J.: Computer-aided pulmonary embolism detection using a novel vessel-aligned multi-planar image representation and convolutional neural networks. In: Navab, N., Hornegger, J., Wells, W.M., Frangi, A.F. (eds.) MICCAI 2015. LNCS, vol. 9350, pp. 62–69. Springer, Cham (2015). https://doi.org/10.1007/978-3-319-24571-3_8

18. Tao, X., Li, Y., Zhou, W., Ma, K., Zheng, Y.: Revisiting Rubik's cube: self-supervised learning with volume-wise transformation for 3D medical image segmentation. arXiv preprint arXiv:2007.08826 (2020)

19. Wu, Z., Xiong, Y., Yu, S.X., Lin, D.: Unsupervised feature learning via non-parametric instance discrimination. In: Proceedings of the IEEE Conference on Computer Vision and Pattern Recognition, pp. 3733–3742 (2018)

20. Zhou, Z., et al.: Models genesis: generic autodidactic models for 3D medical image analysis. In: Shen, D., et al. (eds.) MICCAI 2019. LNCS, vol. 11767, pp. 384–393. Springer, Cham (2019). https://doi.org/10.1007/978-3-030-32251-9_42

Cross-View Label Transfer in Knee MR Segmentation Using Iterative Context Learning

Tong Li[1,3], Kai Xuan[2], Zhong Xue[3], Lei Chen[3], Lichi Zhang[2(✉)], and Dahong Qian[4(✉)]

[1] School of Biomedical Engineering, Shanghai Jiao Tong University, Shanghai, China
[2] Institute for Medical Imaging Technology, School of Biomedical Engineering, Shanghai Jiao Tong University, Shanghai, China
lichizhang@sjtu.edu.cn
[3] Shanghai United Imaging Intelligence Co., Ltd., Shanghai, China
[4] Institute of Medical Robotics, Shanghai Jiao Tong University, Shanghai, China
dahong.qian@sjtu.edu.cn

Abstract. MR images of knee joint are usually collected in axial, coronal, and sagittal views with large slice spacing for clinical study. Current methods either segment images in different views separately or apply super-resolution fusion before 3D segmentation. Knee images segmentation transfer between different views is still an open problem. Moreover, the majority of manual labelling works focus on the sagittal-view, and practically it is hard to collect label maps for the coronal- and axial-views, which are also invaluable for observing knee injuries. In this paper, we propose a novel algorithm to transfer sagittal-view annotations to the other views. First, we build a supervised low-resolution segmentation (LR-Seg) module based on the down-sampled sagittal-view slices to obtain the label map on the target view. And then a context transfer module is proposed to refine the segmentations using target-view context. Then by iterative learning of these two modules, the context from one result can be used to guide the training of the other. Experimental results show that our algorithm can greatly alleviate the burden of manually labeling works from clinicians and gain comparable segmentation results on axial and coronal views.

Keywords: Knee MR images · Multi-view segmentation · Label transfer · Iterative context learning

1 Introduction

Magnetic resonance (MR) imaging is the standard technique for knee disorder diagnosis. In clinical practice, the acquisition of isotropic 3D MR images is impractical due to limited scanning time [1,2]. Therefore, 2D images from three views, namely axial, coronal, and sagittal images are commonly acquired. These images generally have large slice spacing, and clinicians assess the structures of the knee from three different perspectives and make diagnosis comprehensively. Among the three views, sagittal view can be used to assess most of knee injuries, and axial and coronal views can be used in conjunction for reading. For example, torn medial collateral ligament can be observed on coronal view, and axial view is the best view to examine the appearance of the patella cartilage.

Knee segmentation is an essential tool for quantitative and morphological analysis in widely used computer-assisted diagnosis of diseases such as osteoarthritis [3], meniscus injury [4]. However, most methods only focus on the sagittal view [5]. It is not consistent to the assessment procedure where the morphology and function of the knee are evaluated across multiple views in clinical setting. Therefore, it is necessary to develop a 3D segmentation to assist clinical evaluation of the knee. There are some researches on joint multi-view learning [6–8]. But they all need multi-view supervision to infer 3D shape prior and multi-view relationship. Joint multi-view segmentation based on one-view supervision has not been studied extensively. In addition, annotations from multi-views images are generally not available in practice. These annotations are normally done in sagittal view, but the number of the annotations in axial and coronal view is few. It is also time-consuming to annotate the segmentation masks from different views. Fortunately considering the spatial relationship among three views, it is possible to transfer the segmentation mask from the sagittal view to the axial and coronal views in an unsupervised manner.

On the other hand, one solution for 3D segmentation is to first perform super-resolution (SR) to reconstruct 3D images and then apply 3D segmentation algorithm. In general, SR methods can either use shape models [9] or neural network learning [10,11] for supervised or unsupervised isotropic high-resolution image reconstruction. However, the reconstructed images and subsequent tasks such as segmentation based on them, need further verification in clinical. One possible problem is that incorrect reconstruction or artifacts may occur on SR images. What's more, the quality of isotropic SR images obtained is not clear enough given that the in-plane resolution is ~10 times higher than the cross-plane resolution. Finally, compared to annotating in thick-slice sagittal view, 3D annotation is even more difficult.

It is worth noting that multi-view segmentation from single-view annotation belongs to weakly or partially supervised training [12], which aims at using limited segmentation labels to infer the entire label maps. Traditional methods like conditional random fields (CRFs) and some learning-based methods [13, 14] can handle such kind of partially labelled segmentation tasks, but they are usually trained based on expectation-maximization (EM) models or self-training paradigm, which are optimized iteratively and could be sensitive to initial masks. Our method also adopts self-training and iterative optimization paradigm, but it doesn't only focus on the optimization of one view. Source and target views are interactively optimized and provide complementary representation for each other.

In this paper, we propose a novel framework to transfer source-view segmentation label map to the other views. The basic idea is that suppose only the annotations for sagittal-view are available, a supervised low-resolution segmentation (LR-Seg) is trained to obtain initial coarse segmentation by converting other target views into sagittal-view slices. Then, context transfer module is applied to refine the segmentations using the target-view image context. These two modules jointly and iteratively train the two-view (source and target view) segmentation networks in a partially supervised fashion. Our key contributions are building iteratively training the multi-view segmentation by using the LR-Seg module and context transfer module. This algorithm is different from the methods mentioned above. By considering the spatial relationship of multi-view images, reliable initial masks are generated using neural networks, and with context transfer segmentation from different views are refined by iteratively applying cross-view constraints. Experimental results on MR knee image segmentation have confirmed the satisfactory performance of multi-view segmentation with single-view annotation training.

2   Method

The proposed method is organized in three folds: 1) A low-resolution segmentation (LR-Seg) module aims to obtain initial coarse label maps of target views (axial or coronal views) from the known labels of source view (sagittal view) by segmenting on low-resolution slices. 2) A context transfer module is designed to learn cross-view context relationship including intensity and shape in order to refine the initial label maps. 3) By iteratively training the LR-Seg module and context transfer module, they can be gradually optimized. Finally, the trained pipeline can be used to transfer source-view label map to the other two views. Then multi-view segmentation results of MR knee images can be obtained and used to the subsequent diagnostic analysis.

Low-Resolution Segmentation (LR-Seg) Module. Obviously, the supervised segmentation model trained on 2D sagittal slices cannot be performed directly on coronal and axial views whose sagittal images suffer from low-resolution along respective y or z axes in world coordinate system. We thus down-sample the sagittal images and annotations in sagittal view to imitate the resolution in axial and coronal views.

The pipeline of the LR-Seg module is illustrated in Fig. 1, which uses coronal view as target view for example. In initial training stage, only training data from sagittal view are used. LR-Seg modules for two target views are trained based on down-sampled sagittal-view data and annotations. In later iterative stage, target-view data and the label maps predicted from context transfer module would be added into training data. We denote the source view as $x^s = (x_{s1}^s, x_{s2}^s, ..., x_{sn}^s)$, target view as $x^t = (x_{t1}^t, x_{t2}^t, ..., x_{tp}^t)$, and ground truth of source view as $y^s = (y_{s1}^s, y_{s2}^s, ..., y_{sn}^s)$, with x_{si}^s, x_{ti}^t and y_{si}^s representing the high resolution slice i of each volume. Superscripts s or t denote the scanning view of the 3D volume.

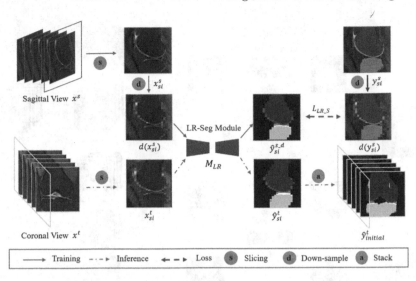

Fig. 1. The pipeline of proposed LR-Seg module (take sagittal view as source view, coronal view as target view.

In subscript si for example, s denotes the direction of slicing, and i denotes the slice number. In initial training stage, the output of LR-Seg module M_{LR} is presented as $\hat{y}_{si}^{s\_d} = M_{LR}(d(x_{si}^s)); \theta_{LR})$, which d refers to down-sampling operation. All the segmentation modules in this paper use combined cross-entropy loss and dice loss during training, which is simply described as:

$$L_{LR}^S = F(\hat{y}^{s\_d}, d(y^s)) = L_{ce}(\hat{y}^{s\_d}, d(y^s))) + L_{dice}(\hat{y}^{s\_d}, d(y^s)) \qquad (1)$$

Initial coarse label map of target views could be obtained by inference of LR-Seg module. Target view x^t can be slicing as $x^t = (x_{s1}^t, x_{s2}^t, ..., x_{sm}^t)$, with x_{sj}^t representing the low-resolution sagittal image. The label prediction of x_{sj}^t is presented as $\hat{y}_{sj}^t = M_{LR}(x_{sj}^t; \theta_{LR})$. The output segmentation maps are stacked into 3D volume $\hat{y}_{initial}^t$ and re-sliced into original high-resolution axial and coronal planes which could show anatomical details:

$$\hat{y}_{initial}^t = (\hat{y}_{s1}^t, ..., \hat{y}_{sm}^t) = (\hat{y}_{t1}^t, ..., \hat{y}_{tp}^t) \qquad (2)$$

The framework of LR-Seg module shown in Fig. 1 is anisotropic 2D U-Net for segmentation, which is similar to vanilla 2D U-Net [15]. The differences lie in the down-sampling and up-sampling layers which are anisotropic for different axes.

Context Transfer Module. By LR-Seg module, the initial segmentation label map on target view can be obtained. But it only roughly learns features on the low-resolution sagittal images. Then context transfer module is needed to refine the initial label map using target-view image context including features of

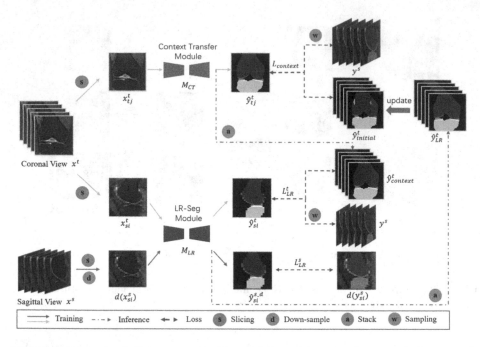

Fig. 2. Framework of the context transfer module and iterative learning (take coronal view as target view for example).

intensity and image shape continuity. As shown in Fig. 2, this module uses high-resolution images of target view as input, which is denoted as x_{tj}^t. The initial label map $\hat{y}_{initial}^t$ is assumed as pseudo ground truth. Considering that it is not reliable, we add the partial supervision from labels of intersective pixels between source and target views. It could only be applied as a supplementary supervision and unable to be independently used as the common pixels are shown as discrete lines on cross-view images and provide insufficient information for optimization. Herein, the vanilla 2D U-Net is adopted. The loss function is defined in Eq. 3. w^s, w^t are operations of sampling common pixels from the source and target views. For inference, the target-view label map $\hat{y}_{t\_context}$ predicted by this module is used for next LR-Seg training.

$$L_{context} = F(\hat{y}^t, \hat{y}_{initial}^t) + F(w^t(\hat{y}^t), w^s(y^s)) \tag{3}$$

Context Transfer Module is proposed by considering the following two reasons. First, according to our observations, the source view can't totally cover the other two views in anatomical space and some sagittal images sliced from the target views may not be included in the training set. Therefore, re-slicing can result in distribution gaps between the training and the testing datasets. Second, the segmentation model possibly produces poor segmentation results at the peripheral slices than the central slices due to more complex shape and appearance variability. Similar problems have been reported in [6]. Clearly, these challenges cannot be solved by using the only supervision from sagittal view. A reasonable way to overcome this obstacles is to incorporate multi-view context and constrain the label transfer using target-view intensity and shape features.

Iterative Learning. It needs to be noticed that the refining procedure from context transfer module depends on the prediction of LR-Seg module. Unreliable initial target-view ground truth in context transfer module can cause possible degenerative optimization. Thus we propose a paradigm named two-view iterative learning to reduce this possible degeneration and jointly learn the context of different views. An iterative cycle contains one training from LR-Seg module and one training from context transfer module respectively. After the first refinement from context transfer module, we add training samples from target view and their prediction as ground truth to LR-Seg training in next cycle. It aims to let LR-Seg module learn from the target view to improve performance on peripheral slices which could be missing in source view but exist in target view. In addition, LR-Seg module is strictly supervised by original training samples with the real annotations. As shown in Fig. 2, LR-Seg module in iterative stage is supervised by three folds: target-view updated label map $\hat{y}^t_{context}$, common ground truth $w^s(y^s)$ and source-view ground truth y^s. y^s and $w^s(y^s)$ are used to ensure the model trained in the direction of positive optimization. $\hat{y}^t_{context}$ is used to guide LR-Seg to learn from the knowledge distilled from target-view. The loss function can be defined as:

$$
\begin{aligned}
L_{LR} &= L^t_{LR} + L^s_{LR} \\
&= F(\hat{y}^t, \hat{y}^t_{context}) + F(w^t(\hat{y}^t), w^s(y^s)) + F(\hat{y}^{s\_d}, d(y^s))
\end{aligned}
\tag{4}
$$

The output of source view and target view are denoted as $\hat{y}^s_{LR}$ and $\hat{y}^t_{LR}$ by LR-Seg inference. For the next training of context transfer module, $\hat{y}^t_{initial}$ is updated to $\hat{y}^t_{LR}$ as pseudo ground truth. As two sub-modules are learning from source view and target view alternately, the prediction gradually approaches the true label of target view.

3   Experiments

Dataset. We have collected a private dataset of knee joint MR images from 160 subjects. For each subject, PD fat suppression MR sequences are scanned along axial, coronal and sagittal directions. It contains 58.8% female subjects and 41.2% male subjects. The age of subjects is from 16 to 64. Nearly 2.5% subjects have been diagnosed as osteoarthritis, and 46.25% subjects have been diagnosed as non-osteoarthritis. The voxel spacing of sagittal view is 0.25 mm × 0.25 mm × 3.3 mm. The voxel spacing of coronal and axial views is 0.306 mm × 0.306 mm × 4 mm. The average slice number of three views is 19 when unifying them into the same slice thickness. All sagittal-view images have been annotated by radiologists, and 26 subjects have coronal-view labels which are used to evaluate the segmentation performance of the model. Dice Similarity Coefficient (Dice) is used to measure the pixel-wise consistency between the predicted label map and the ground truth. 26 subjects on coronal view are used for evaluation.

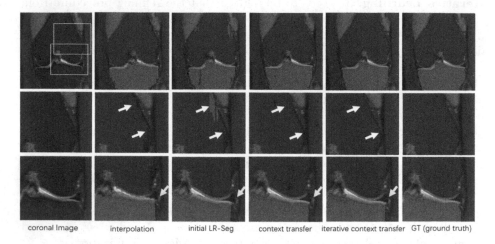

coronal Image   interpolation   initial LR-Seg   context transfer   iterative context transfer   GT (ground truth)

Fig. 3. Results on coronal view. Top: from left to right, original coronal image, predicted label map from straightly applying nearest interpolation (interpolation), the first LR-Seg module training (initial LR-Seg), the first context transfer training (context transfer), the second context transfer training (iterative context transfer) and ground truth (GT). Middle and bottom: the details in two bounding-boxes annotated in the first row.

Experiments Settings and Results. In training stage, we cropped roughly the same anatomical space of three views. Input size is 224×18 in LR-Seg module and 224×224 in context transfer module. The learning rate for the iterative two modules was 3×10^{-4} and adam optimizer was adopted with $\beta_1 = 0.9$ and $\beta_2 = 0.999$ in training stage. We present visualization results on coronal and axial views in Fig. 3 and Fig. 4, and numerical results on coronal view in Table 1. Besides evaluating the performance of different stages in our proposed method, we also compare with the following methods: 1) apply nearest interpolation (Nearest Interp.) to transfer sagittal-view label to the target view. 2) cross view segmentation (Cross-view Seg.). Though the resolutions of source and target views are different in some direction, 3D segmentation model trained on one view has potentials to make inference on the other views with corresponding input. 3) segmentation over a 3D tri-linear interpolated MR volume (Interp. Seg). It could be reasonable as the label map of interpolated 3D volume can provide smooth and consistent shape of any view. We use V-Net [16] as the 3D segmentation network.

Figure 3 illustrates the results of coronal view. We iteratively train 2 cycles which train context transfer module twice and LR-Seg module twice. Compared to the results of nearest interpolation, the boundary of initial LR-Seg label map gets smoother. After iterative context learning of two modules, the label map has been refined a lot. Under and incorrect segmentation areas have been corrected. In the first bounding-box in the second row, as the arrow shows, the wrongly labeled holes in the initial LR-Seg Module's output have been filled. The mask of femur bone has gradually extended to the boundary. In the second bounding-box in the last row, the segmentation area of meniscus gets more and more accurate. The results have proved the validity of our framework.

Table 1. Comparison of baselines and our proposed method evaluated on the coronal view in terms of the mean and the standard deviation of Dice score ($dice \pm std$, %). The compared labels include femur bones (FB) and tibial bones (TB), femoral cartilage (FC), tibial cartilage (TC) and meniscus.

Method	FB	TB	FC	TC	Meniscus
Nearest Interp.	92.5 ± 2.4	93.2 ± 3.5	66.7 ± 3.9	70.1 ± 2.9	74.5 ± 4.4
Cross-view Seg.	94.2 ± 1.9	94.8 ± 1.6	56.9 ± 4.4	78.7 ± 3.3	83.6 ± 2.6
Interp. Seg.	92.8 ± 2.3	93.1 ± 3.6	69.8 ± 4.2	74.0 ± 3.7	74.8 ± 5.0
LR-Seg	94.8 ± 2.8	94.8 ± 2.6	79.4 ± 3.8	83.0 ± 4.1	88.1 ± 3.8
Context Trans.	97.2 ± 2.7	97.1 ± 1.2	86.1 ± 3.8	89.7 ± 3.1	93.7 ± 2.7
Iterative Context Trans.	$\mathbf{98.0 \pm 2.5}$	$\mathbf{98.0 \pm 1.4}$	$\mathbf{88.8 \pm 4.6}$	$\mathbf{91.7 \pm 3.6}$	$\mathbf{97.9 \pm 3.6}$

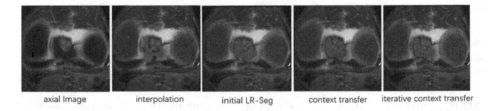

axial Image   interpolation   initial LR-Seg   context transfer   iterative context transfer

Fig. 4. Segmentation results of axial view.

As shown in Table 1, we compare the proposed method with three baselines mentioned previously. Results of different stages in our method are from the first training of LR-Seg, the first training of context transfer (Context Trans.), and the second training of context transfer (Iterative Context Trans.) in the next iterative cycle. All results from our proposed method achieve higher than the three baselines. The cross-view segmentation could cause poor results as the discrepancy from image resolution and appearance between different views can't be ignored in 3D segmentation model. The performance based on tri-linear interpolated volume is even worse than the cross-view segmentation. Although some advanced super-resolution methods may produce high quality images, they are harder to train and may need high-resolution image as supervision. Segmentation results would largely rely on the synthesis stage. More importantly, multi-view segmentation based on super-resolution method would complicate the problem. Therefore, we try to simplify it by using the relationship of multi views and directly make segmentation. Compared with the results of initial LR-Seg module, Dice score of all classes after the first training of context transfer have been improved a lot especially in the femoral cartilage (FC) and tibial cartilage (TC). Then iterative learning could improve dice score by 2%–4.2% and get the best performance.

Figure 4 shows the results of axial view by the proposed method. The training process is the same as coronal view. The initial label map from LR-Seg module has some wrong and inconsecutive labels. On the both sides of the image, some areas are wrong labeled as background. After introducing context from axial view, the label map has been improved a lot and the segmentation boundary gets smoother and more continuous. Some wrong labels has been cleared.

4   Conclusion

In this work, we have presented a framework to transfer one-view segmentation labels onto another view when no ground truth is available. Two modules are designed to achieve this goal: LR-Seg module for low-resolution sagittal image segmentation, and context transfer module for refining the segmentation by learning the image context across different views. By iterative training, the proposed model can learn complementary features from different views and improve the segmentation prediction on the target view. The proposed method has been validated on coronal and axial views of MR knee images. We show visualization results on both target views and numeric result on coronal view. The method can be applied to other segmentation transfer tasks any two between multiple views and decrease the workload of manually labeling works for MR image segmentation.

Acknowledgement. This work was supported in part by the National Key Research and Development Program of China under Grant 2018YFC0116402, Department of Science and Technology of Zhejiang Province Key Research and Development Program under Grant 2017C03029, Shanghai Pujiang Program (19PJ1406800), and Interdisciplinary Program of Shanghai Jiao Tong University.

References

1. Nacey, N.C., Geeslin, M.G., Miller, G.W., Pierce, J.L.: Magnetic resonance imaging of the knee: an overview and update of conventional and state of the art imaging. J. Magn. Reson. Imaging JMRI **45**(5), 1257–1275 (2017)
2. Naraghi, A.M., White, L.M.: Imaging of athletic injuries of knee ligaments and menisci: sports imaging series. Radiology **281**(1), 23–40 (2016)
3. Ambellan, F., Tack, A., Ehlke, M., Zachow, S.: Automated segmentation of knee bone and cartilage combining statistical shape knowledge and convolutional neural networks: data from the osteoarthritis initiative. Med. Image Anal. **52**, 109–118 (2019)

4. Pedoia, V., Norman, B., Mehany, S.N., Bucknor, M.D.: 3D convolutional neural networks for detection and severity staging of meniscus and PFJ cartilage morphological degenerative changes in osteoarthritis and anterior cruciate ligament subjects. J. Magn. Reson. Imaging **49**, 400–410 (2019)
5. Liu, Q., et al.: Multi-class gradient harmonized dice loss with application to knee MR image segmentation. In: Shen, D., et al. (eds.) MICCAI 2019. LNCS, vol. 11769, pp. 86–94. Springer, Cham (2019). https://doi.org/10.1007/978-3-030-32226-7_10
6. Chen, C., Biffi, C., Tarroni, G., Petersen, S., Bai, W., Rueckert, D.: Learning shape priors for robust cardiac MR segmentation from multi-view images. In: Shen, D., et al. (eds.) MICCAI 2019. LNCS, vol. 11765, pp. 523–531. Springer, Cham (2019). https://doi.org/10.1007/978-3-030-32245-8_58
7. Mortazi, A., Karim, R., Rhode, K., Burt, J., Bagci, U.: *CardiacNET*: segmentation of left atrium and proximal pulmonary veins from MRI using multi-view CNN. In: Descoteaux, M., Maier-Hein, L., Franz, A., Jannin, P., Collins, D.L., Duchesne, S. (eds.) MICCAI 2017. LNCS, vol. 10434, pp. 377–385. Springer, Cham (2017). https://doi.org/10.1007/978-3-319-66185-8_43
8. Setio, A.A.A., et al.: Pulmonary nodule detection in CT images: false positive reduction using multi-view convolutional networks. IEEE Trans. Med. Imaging **35**(5), 1160–1169 (2016)
9. Sui, Y., Afacan, O., Gholipour, A., Warfield, S.K.: Isotropic MRI super-resolution reconstruction with multi-scale gradient field prior. In: Shen, D., et al. (eds.) MICCAI 2019. LNCS, vol. 11766, pp. 3–11. Springer, Cham (2019). https://doi.org/10.1007/978-3-030-32248-9_1
10. Chaudhari, A.S., et al.: Super-resolution musculoskeletal MRI using deep learning. Magn. Reson. Med. **80**(5), 2139–2154 (2018)
11. Xuan, K., et al.: Reconstruction of isotropic high-resolution MR image from multiple anisotropic scans using sparse fidelity loss and adversarial regularization. In: Shen, D., et al. (eds.) MICCAI 2019. LNCS, vol. 11766, pp. 65–73. Springer, Cham (2019). https://doi.org/10.1007/978-3-030-32248-9_8
12. Papandreou, G., Chen, L.C., Murphy, K.P., Yuille, A.L.: Weakly-and semi-supervised learning of a deep convolutional network for semantic image segmentation. In: Proceedings of the IEEE International Conference on Computer Vision, pp. 1742–1750 (2015)
13. Rajchl, M., et al.: DeepCut: object segmentation from bounding box annotations using convolutional neural networks. IEEE Trans. Medical Imaging **36**, 674–683 (2017)
14. Wang, G., et al.: Interactive medical image segmentation using deep learning with image-specific fine-tuning. IEEE Trans. Med. Imaging **37**, 1562–1573 (2018)
15. Ronneberger, O., Fischer, P., Brox, T.: U-Net: convolutional networks for biomedical image segmentation. In: Navab, N., Hornegger, J., Wells, W.M., Frangi, A.F. (eds.) MICCAI 2015. LNCS, vol. 9351, pp. 234–241. Springer, Cham (2015). https://doi.org/10.1007/978-3-319-24574-4_28
16. Milletari, F., Navab, N., Ahmadi, S.: V-net: fully convolutional neural networks for volumetric medical image segmentation. In: 2016 Fourth International Conference on 3D Vision (3DV), pp. 565–571 (2016)

Continual Class Incremental Learning for CT Thoracic Segmentation

Abdelrahman Elskhawy[1,2]($\boxtimes$), Aneta Lisowska[2], Matthias Keicher[1],
Joseph Henry[2], Paul Thomson[2], and Nassir Navab[1,3]

[1] Computer Aided Medical Procedures, Technische Universität München,
Munich, Germany
{a.elskhawy,matthias.keicher}@tum.de
[2] Canon Medical Research Europe Ltd., Edinburgh, UK
{Aneta.Lisowska,Joseph.Henry,Paul.Thomson}@eu.medical.canon
[3] Computer Aided Medical Procedures, Johns Hopkins University, Baltimore, USA
navab@cs.tum.edu

Abstract. Deep learning organ segmentation approaches require large
amounts of annotated training data, which is limited in supply due to reasons of confidentiality and the time required for expert manual annotation. Therefore, being able to train models incrementally without having
access to previously used data is desirable. A common form of sequential
training is fine tuning (FT). In this setting, a model learns a new task
effectively, but loses performance on previously learned tasks. The Learning without Forgetting (LwF) approach addresses this issue via replaying
its own prediction for past tasks during model training. In this work, we
evaluate FT and LwF for class incremental learning in multi-organ segmentation using the publicly available AAPM dataset. We show that
LwF can successfully retain knowledge on previous segmentations, however, its ability to learn a new class decreases with the addition of each
class. To address this problem we propose an adversarial continual learning segmentation approach (ACLSeg), which disentangles feature space
into task-specific and task-invariant features. This enables preservation
of performance on past tasks and effective acquisition of new knowledge.

Keywords: Continual learning · CT segmentation · Adversarial
learning · Latent space factorisation · Incremental class learning

1 Introduction

The best performing deep learning solutions are trained on a large number of
annotated training examples. However, it might be infeasible to annotate large
amounts of data to train specialised models from scratch for each new medical

Electronic supplementary material The online version of this chapter (https://
doi.org/10.1007/978-3-030-60548-3_11) contains supplementary material, which is
available to authorized users.

S. Albarqouni et al. (Eds.): DART 2020/DCL 2020, LNCS 12444, pp. 106–116, 2020.
https://doi.org/10.1007/978-3-030-60548-3_11

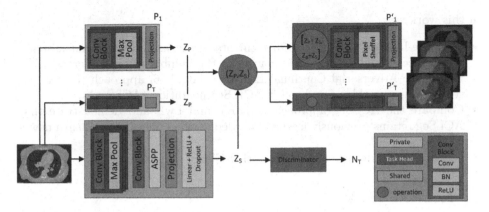

Fig. 1. ACLSeg architecture. Considering a sequence of T tasks, a private module (P) and a task head (P') are added for each task, while the shared module (S) is common between all tasks. Both S and the P for each task receive the input CT slice and generate the shared (Z_S) and private (Z_P) embeddings respectively. For each task head, Z_S and Z_P are multiplied and added element-wise, then concatenated to be further processed using Conv block and PixelShuffle modules to generate the final output. The discriminator receives Z_S only and tries to predict the task label N_T in an adversarial *minmax* game with the shared module. A projection layer towards the end of each module reduces the number of channels to 1.

imaging problem. This can be due to privacy regulations that impose constraints on sharing patients' sensitive data, fragmented healthcare systems, and/or the time and expense required for expert manual annotation. Therefore, being able to train models incrementally without having access to previously used data is desirable.

The most common form of sequential training is fine tuning (FT). In this setting, a pre-trained model can learn a new task effectively from a smaller amount of data, but at the cost of losing its ability to perform previously learned tasks; a phenomenon known as catastrophic forgetting [18]. Continual learning (CL) approaches intend to address this issue either via structural growth such as [24,31], which relies on adding task-specific modules, regularisation-based methods such as [2,11], which penalise significant changes to the previous tasks' representations, or replay-based methods such as [22,27] which replay previous data either explicitly or via pseudo-rehearsal. For the purposes of this work, we do not discuss explicit replay-based methods as we assume that direct access to previous data is not possible. Learning without Forgetting (LwF) [17], which combines both regularisation-based and pseudo-rehearsal techniques, has been the state-of-the-art medical imaging continual learning method for situations in which access to previous training data is not possible. It has shown promising results in medical imaging for both incremental domain learning [15] and incremental class learning [19]. However, it has not been previously evaluated on the incremental class learning problem with a task sequence exceeding two tasks.

In this work we:

- Evaluate LwF on a sequence of 5 segmentation tasks and show that it struggles to accommodate more information as the number of tasks increases.
- Adopt the Adversarial Continual Learning (ACL) [6] approach to work for segmentation problems and call it ACL Segmentation (ACLSeg)
- Compare the ACLSeg approach with FT and LwF and demonstrate that ACLSeg retains previously learned knowledge while being able to learn newly added tasks.
- Explore task-order robustness for both ACLSeg and LwF.

2   Related Work

Continual Learning in the Medical Domain. Although there are various CL approaches proposed for natural image classification tasks [21], not many of them have been applied to medical image segmentation. For domain incremental learning, Ozgung et al. [20] proposed learning rate regularisation to Memory Aware Synapses [1] to perform MRI brain segmentation. Lenga et al. have shown that LwF outperforms elastic weight consolidation (EwC) [11], when applied to incremental X-ray domain learning [15].

In the incremental class learning setting, Baweja et al. [2] used EWC to sequentially learn cerebrospinal fluid segmentation followed by grey and white matter segmentation tasks. Ozdemir and Goksel [19] applied Learning without Forgetting (LwF) to sequential learning of tibia and femur bone in MRI of the knee. The authors suggested that LwF is a viable CL solution when sharing patient data is not possible due to privacy concerns. When retention of representative samples from past datasets is possible, the authors suggested to use AeiSeg, which extends LwF via sample replay, leading to improved knowledge preservation. In this work we assume that there is no access to previous data, therefore we do not include AeiSeg in our comparison.

Latent Space Disentanglement. Multi-view learning [16] exploits different modalities of the data to maximise the performance. For class incremental learning, factorising the data representation into both shared and task-specific parts helps prevent forgetting. While the learned shared representation is less susceptible to forgetting, as it is task-invariant, preventing forgetting in the private representations can be achieved by using small sub-modules per class that are frozen upon finishing learning that specific class. Latent space factorisation can be achieved by either Adversarial training as in [6], orthogonality constraints as in [25], or both to ensure complete enforced factorisation.

3   ACL Segmentation (ACLSeg)

Our objective is to learn to segment T organs in CT scans in a sequential manner. To achieve this, we build upon the ACL approach [6], initially developed for incremental classification problems on MNiST [14] and CIFAR [12], and adopt it to solve segmentation problems. Figure 1 shows the architecture for ACLSeg including the modifications that are described in this section.

Consider a sequence of T tasks to be learned one task at a time. For the very first task, the model consists of the shared module, the discriminator, one private module P_1, and one task head P_1'. The discriminator attempts to predict the task label in a *minmax* game with the shared module. When adding new segmentation tasks we add a task head and a private module, while the shared module remains common to all tasks.

The main idea of ACL is to learn a disjoint latent space representation composed of task-invariant (shared) latent space, represented by Z_S, and task-specific (private) latent space, represented by Z_P. A task-specific head receives both Z_S and Z_P to generate the final output. The objective function for ACL is:

$$\mathcal{L}_{\text{ACLSeg}} = \lambda_1 \mathcal{L}_{\text{task}} + \lambda_2 \mathcal{L}_{\text{adv}} + \lambda_3 \mathcal{L}_{\text{diff}} \qquad (1)$$

Where $\mathcal{L}_{\text{task}}$ is the task loss (Binary Cross Entropy loss is used for each segmentation task), $\mathcal{L}_{\text{adv}}$ is the T-way classification cross-entropy adversarial loss, and $\mathcal{L}_{\text{diff}}$ is an orthogonality constraint introduced in [25], also known as the difference loss [3] in domain adaptation literature. $\mathcal{L}_{\text{diff}}$ ensures further factorisation of the shared and private features. $\mathcal{L}_{\text{adv}}$ and $\mathcal{L}_{\text{diff}}$ are described in detail in the supplementary materials and [6]. λ_1, λ_2, and λ_3 are regularisers to control the strength of each loss component.

Adaptation to CT Segmentation. To adapt ACL to segmentation of CT slices, we introduce a few changes to the different components of the original architecture, taking into consideration the final model size.

Shared Module. To enrich the extracted features in the shared module, we used an encoder-based module with Atrous Spatial Pyramid Pooling (ASPP) [5]. ASPP allows explicit control over the resolution of extracted features and adjusts a filter's field of view to capture multi-context information at reduced computational cost. This is desirable due to the size of medical data, and the need to segment small anatomical structures.

Task Heads. We introduce two modifications to the task heads: **A)** While the original ACL paper proposes concatenating Z_S and Z_P to generate the final output, [8] suggested that this is not the optimal way to fuse multiple streams of information. In order to ensure enriched representations in each task head, we replace the concatenation operation with additive and multiplicative counter parts, i.e. Z_S and Z_P are multiplied and added, then concatenated over the channel dimension and tehn passed to the respective task head. Given the two vectors Z_S and Z_P of length *Latent_dim* each, the input to the respective task head is the concatenation of both $(Z_S \odot Z_P)$ and $(Z_S \oplus Z_P)$ along the channel dimension resulting in a two-channel feature map of final size *Latent_dim* $\times 2$. **B)** We designed the task head to be compact to make the model scalable as we add a task head for each newly added task. To achieve this, we upsample the private and shared embeddings in two stages with a 4X upsampling factor at each stage. Unlike the U-Net architecture [23], where multi-scale features are fused via skip connections from the encoder to the decoder, our model architecture relies on the output from the embedding vectors only. To improve over the upsampling and to be able to recover some of the segmentation details with such a large upsampling factor, we replace the Convolution Transpose upsampling with sub-pixel convolutions proposed in [26]. This approach uses regular convolution layers followed by a Phase Shift reshaping operation which solves the checkerboard artifacts and improves the segmentation results as shown in [4,7,13]. For further details on how each of these modifications contributed to the segmentation performance please refer to Table 1 in the Supplementary Material.

3.1   Evaluation Metrics

In order to assess a CL system, the system needs to be evaluated on two different aspects. First, segmentation quality, which is useful for tracking the running segmentation score to ensure that the model is providing meaningful segmentation. We report this as the Dice Coefficient (DC). Second, knowledge retention and the ability to learn new information. For this, we adopt the metrics proposed in [9] with slight modifications to Ω_{new} calculations, in which we normalise the value to fall in the range $[0, 1]$. This makes the three Ω values comparable across different continual learning approaches. Therefore, our modified knowledge retention metrics are:

$$\Omega_{base} = \frac{1}{T-1} \sum_{i=2}^{T} \frac{\alpha_{base,i}}{\alpha_{ideal,base}} \tag{2}$$

$$\Omega_{new} = \frac{1}{T-1} \sum_{i=2}^{T} \frac{\alpha_{new,i}}{\alpha_{ideal,i}} \tag{3}$$

$$\Omega_{all} = \frac{1}{T-1} \sum_{i=2}^{T} \frac{\overline{\alpha_{all,0:i}}}{\overline{\alpha_{ideal,0:i}}} \tag{4}$$

where T is the total number of classes, $\alpha_{base,i}$ is the DC of the first class after i classes have been learned, $\alpha_{new,i}$ is the DC of class i immediately after it is learned, $\overline{\alpha_{all,0:i}}$ is the mean DC of all the classes that have been seen so far up to and including step i, $\overline{\alpha_{ideal,0:i}}$ is the offline mean DC of all the classes that have been seen so far up to and including step i, by jointly training the model on all the available data at once, and $\alpha_{ideal,base}$, and $\alpha_{ideal,i}$ are the offline ideal DC of the base and the ith class respectively.

Ω_{base} measures the model retention of the first learned class after learning subsequent classes, Ω_{new} measures the model ability to learn new classes, and Ω_{all} computes how well a model can both retain prior knowledge and acquire new information. All Ω values $\in [0, 1]$ unless a CL model exceeds the upper

Table 1. Ω scores, (Std. dev. of 3 runs), and overall dice score of the final model for class incremental learning on 5 classes

	Ω_{base}	Ω_{new}	Ω_{all}	Overall dice score
FT	0.03(0.000)	0.93(0.090)	0.33(0.020)	0.14(0.008)
LwF	**1.09(0.010)**	0.82(0.005)	0.96(0.026)	0.75(0.010)
ACLSeg	1.00(0.005)	**0.96(0.006)**	**0.99(0.004)**	**0.80(0.005)**

bound. Since the $\alpha_{ideal,n}$ is obtained from offline training the same model on all the data at once, the architectural choice of the model does not affect our comparison.

4   Experiments and Results

4.1   Experimental Setup

Datasets. We experiment with the publicly available AAPM Thoracic auto-segmentation challenge dataset (AAPM) [29]. The AAPM dataset has segmentations for 5 organs: spinal cord, right lung, left lung, heart, and oesophagus. The training set is composed of 30 scans, which are further split into 5 subsets, one for each class, 6 validation and 24 testing CT scans.

Training Scheme and Hyperparameters. For all experiments, the models were trained to convergence using EarlyStopping on a validation dataset. We chose an initial learning rate (lr) of 1e-3, which is reduced with a factor of 3 on validation loss plateau. The algorithm is implemented using Pytorch, and we train the network using the Adam optimiser [10]. The inputs were normalised, and resized to 256×256 instead of 512×512 with no other data augmentation techniques applied. *Latent_dim* is chosen to be 256, and λ_1, λ_2, and λ_3 are 1, 0.05, and 0.3 respectively. Empirically, higher values of λ_2 would render the adversarial training unstable, while higher values of λ_3 would concentrate most of the information in private modules leading to sub-optimal latent space separation. We average all the demonstrated results across three runs to report the mean and standard deviation.

Baselines. We compare ACLSeg with LwF which represents the state-of-the-art in class incremental learning for segmentation problems[19] when access to previous data is not possible. For this purpose, we adopted the Multi-head U-Net structure proposed in [19]. We used Binary Cross Entropy as the segmentation loss for each task, and regressed the class probabilities (logits) of the previous model using mean square loss as our knowledge distillation loss, as proposed in [28]. We also perform naive Fine Tuning (FT) in which a single model is trained sequentially with no forgetting prevention techniques which serves as a lower bound. For the upper bound (ideal), we jointly train the model in a multitask setting with all the available data.

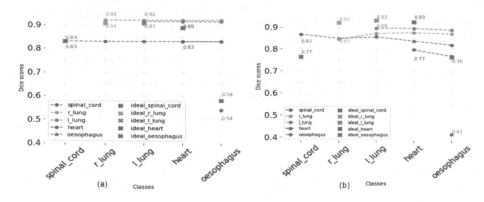

Fig. 2. Dice coefficients for a model trained sequentially on the AAPM dataset along with the corresponding ideal dice coefficients obtained from offline training. (a) ACLSeg, (b) LwF

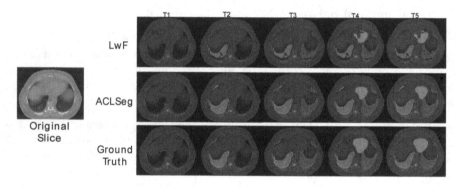

Fig. 3. Ground truth and segmentation results for a given input slice using LwF and ACLSeg after learning each task in OrderA

4.2   Results

5-Split AAPM. We split the training dataset into 5 subsets, one for each of the classes, and learned one class at a time. Table 1 shows the obtained Omega scores along with the overall dice score of the final model. The results are reported on the sequence: Spinal Cord, Right Lung, Left Lung, Heart, and Oesophagus (OrderA). We observe that although LwF is able to retain the performance of the base task, shown by the large Ω_{base} score, it struggles to accommodate new information as more classes are added. Figure 2 shows the change in dice scores after the addition of new classes. We observe that LwF performance on some of the previous classes exhibits a small degradation with the addition of new classes, and does not fully learn the 4th and 5th classes (See Fig. 3). In contrast, the FT approach shows a high Ω_{new} score as it focuses on learning the new classes, at the expense of losing previous information. However Ω_{new} score does not reach the ideal score, possibly due to the differences between subsequent tasks, which gives unfavorable initialisation of the model weights for learning the new task. ACLSeg combines both capabilities by being able to retain a consistent performance on all the previously learned tasks, while having the ability to reach near-ideal performance on subsequent tasks as shown in Fig. 2, and reflected in the 5% increase in the overall dice score of the final model compared to LwF in Table 1. Our interpretation is that this might be due to the disentanglement of the latent space which preserves the task-related knowledge in the respective private module, while being able to update the shared module with only the task-invariant information, hence preventing catastrophic forgetting. Figure 4 shows the t-SNE visualisation of the generated embeddings, which shows that the shared embeddings form a uniform distribution of samples belonging to all classes which can not be uncovered, while the private modules are successful in uncovering class labels in their latent space. We point out that although ACLSeg shows zero forgetting, it struggles to learn the last class (Oesophagus) due to its complexity and severe under-representation in the dataset. This is also true for LwF and ideal training and leaves room for improvement.

(a)   (b)

Fig. 4. T-sne visualisation of the embeddings generated by a) the shared module and b) private modules

Task-Order Robustness. Yoon et al. showed that CL model performance significantly varies based on the order in which the tasks are learned [30]. Since this large variance might cause an issue in the medical domain, we investigate different task orders. We pick two different sequences in addition to OrderA, and report our results in Table 2. OrderB represents the sequence "Oesophagus, Heart, Left Lung, Right Lung, and Spinal Cord" which starts with the hardest-to-segment class. OrderC represents the sequence "Left Lung, Right Lung, Spinal Cord, Heart, and Oesophagus" which starts with an easy-to-segment class followed by a medium difficulty one, while OrderA starts with a medium difficulty class followed by an easy one. From Table 2 we observe that starting with a hard-to-segment task has an effect on the base score as the model was not able to fully learn the base class, Oesophagus in this case, however, it was able to maintain the performance on this class till the end of sequential training (see supplementary material for detailed results).

Table 2. Ω scores and (Std. dev. of 3 runs) for different task orders

	Ω_{base}		Ω_{new}		Ω_{all}	
	ACLSeg	LwF	ACLSeg	LwF	ACLSeg	LwF
OrderA	1.01(0.005)	**1.09(0.010)**	**0.96(0.006)**	0.82(0.005)	**0.99(0.004)**	0.96(0.026)
OrderB	**0.82(0.010)**	0.53(0.140)	**0.98(0.001)**	0.83(0.080)	**0.93(0.004)**	0.70(0.040)
OrderC	0.99(0.002)	**1.0(0.004)**	**0.94(0.005)**	0.84(0.020)	**0.99(0.003)**	0.94(0.009)

5   Conclusion

We adapted an adversarial continual learning approach to medical data (ACLSeg) and evaluated it on an incremental thoracic segmentation problem. We demonstrated that ACLSeg retains knowledge equally as well as LwF, while being able to achieve better performance on newly added tasks. For both approaches task order affects the anatomy segmentation performance, however for ACLSeg the knowledge retention is preserved. We also showed that ACLSeg has a disentangled latent space that is composed of task-invariant and task-specific representations which might be useful for model explainability and privacy preservation.

References

1. Aljundi, R., Babiloni, F., Elhoseiny, M., Rohrbach, M., Tuytelaars, T.: Memory aware synapses: learning what (not) to forget. In: Proceedings of the European Conference on Computer Vision (ECCV), pp. 139–154 (2018)
2. Baweja, C., Glocker, B., Kamnitsas, K.: Towards continual learning in medical imaging. arXiv preprint arXiv:1811.02496 (2018)
3. Bousmalis, K., Trigeorgis, G., Silberman, N., Krishnan, D., Erhan, D.: Domain separation networks. In: Advances in Neural Information Processing Systems, pp. 343–351 (2016)
4. Chen, K., Fu, K., Yan, M., Gao, X., Sun, X., Wei, X.: Semantic segmentation of aerial images with shuffling convolutional neural networks. IEEE Geosci. Remote Sens. Lett. **15**(2), 173–177 (2018)
5. Chen, L.C., Zhu, Y., Papandreou, G., Schroff, F., Adam, H.: Encoder-decoder with atrous separable convolution for semantic image segmentation. In: Proceedings of the European Conference on Computer Vision (ECCV), pp. 801–818 (2018)
6. Ebrahimi, S., Meier, F., Calandra, R., Darrell, T., Rohrbach, M.: Adversarial continual learning. arXiv preprint arXiv:2003.09553 (2020)
7. Gao, H., Yuan, H., Wang, Z., Ji, S.: Pixel deconvolutional networks. arXiv preprint arXiv:1705.06820 (2017)
8. Jayakumar, S.M., et al.: Multiplicative interactions and where to find them. In: International Conference on Learning Representations (2020)
9. Kemker, R., McClure, M., Abitino, A., Hayes, T.L., Kanan, C.: Measuring catastrophic forgetting in neural networks. In: Thirty-Second AAAI Conference on Artificial Intelligence (2018)

10. Kingma, D.P., Ba, J.: Adam: a method for stochastic optimization. arXiv preprint arXiv:1412.6980 (2014)
11. Kirkpatrick, J., et al.: Overcoming catastrophic forgetting in neural networks. Proc. Natl. Acad. Sci. **114**(13), 3521–3526 (2017)
12. Krizhevsky, A.: Learning multiple layers of features from tiny images. Technical report (2009)
13. Lachinov, D.: Segmentation of thoracic organs using pixel shuffle. In: SegTHOR@ ISBI (2019)
14. LeCun, Y., Cortes, C., Burges, C.: Mnist handwritten digit database, no. 2. ATT Labs. http://yann.lecun.com/exdb/mnist (2010)
15. Lenga, M., Schulz, H., Saalbach, A.: Continual learning for domain adaptation in chest x-ray classification. arXiv preprint arXiv:2001.05922 (2020)
16. Li, Y., Yang, M., Zhang, Z.: A survey of multi-view representation learning. IEEE Trans. Knowl. Data Eng. **31**(10), 1863–1883 (2018)
17. Li, Z., Hoiem, D.: Learning without forgetting. IEEE Trans. Pattern Anal. Mach. Intell. **40**(12), 2935–2947 (2017)
18. McCloskey, M., Cohen, N.J.: Catastrophic interference in connectionist networks: the sequential learning problem. In: Psychology of Learning and Motivation, vol. 24, pp. 109–165. Elsevier (1989)
19. Ozdemir, F., Goksel, O.: Extending pretrained segmentation networks with additional anatomical structures. Int. J. Comput. Assist. Radiol. Surg. **14**(7), 1187–1195 (2019). https://doi.org/10.1007/s11548-019-01984-4
20. Özgün, S.Ö., Rickmann, A.M., Roy, A.G., Wachinger, C.: Importance driven continual learning for segmentation across domains. arXiv preprint arXiv:2005.00079 (2020)
21. Parisi, G.I., Kemker, R., Part, J.L., Kanan, C., Wermter, S.: Continual lifelong learning with neural networks: a review. Neural Netw. **113**, 54–71 (2019)
22. Rebuffi, S.A., Kolesnikov, A., Sperl, G., Lampert, C.H.: iCaRL: incremental classifier and representation learning. In: Proceedings of the IEEE Conference on Computer Vision and Pattern Recognition, pp. 2001–2010 (2017)
23. Ronneberger, O., Fischer, P., Brox, T.: U-Net: convolutional networks for biomedical image segmentation. In: Navab, N., Hornegger, J., Wells, W.M., Frangi, A.F. (eds.) MICCAI 2015. LNCS, vol. 9351, pp. 234–241. Springer, Cham (2015). https://doi.org/10.1007/978-3-319-24574-4_28
24. Rusu, A.A., et al.: Progressive neural networks. arXiv preprint arXiv:1606.04671 (2016)
25. Salzmann, M., Ek, C.H., Urtasun, R., Darrell, T.: Factorized orthogonal latent spaces. In: Proceedings of the Thirteenth International Conference on Artificial Intelligence and Statistics, pp. 701–708 (2010)
26. Shi, W., et al.: Real-time single image and video super-resolution using an efficient sub-pixel convolutional neural network. In: Proceedings of the IEEE Conference on Computer Vision and Pattern Recognition, pp. 1874–1883 (2016)
27. Shin, H., Lee, J.K., Kim, J., Kim, J.: Continual learning with deep generative replay. In: Advances in Neural Information Processing Systems, pp. 2990–2999 (2017)
28. Shmelkov, K., Schmid, C., Alahari, K.: Incremental learning of object detectors without catastrophic forgetting. In: Proceedings of the IEEE International Conference on Computer Vision, pp. 3400–3409 (2017)

29. Yang, J., et al.: Data from lung CT segmentation challenge. The cancer imaging archive (2017)
30. Yoon, J., Kim, S., Yang, E., Hwang, S.J.: Scalable and order-robust continual learning with additive parameter decomposition. arXiv preprint arXiv:1902.09432 (2019)
31. Yoon, J., Yang, E., Lee, J., Hwang, S.J.: Lifelong learning with dynamically expandable networks. arXiv preprint arXiv:1708.01547 (2017)

First U-Net Layers Contain More Domain Specific Information Than the Last Ones

Boris Shirokikh[1,2], Ivan Zakazov[1,3(✉)], Alexey Chernyavskiy[3], Irina Fedulova[3], and Mikhail Belyaev[1]

[1] Skolkovo Institute of Science and Technology, Moscow, Russia
Ivan.Zakazov@skoltech.ru
[2] Kharkevich Institute for Information Transmission Problems, Moscow, Russia
[3] Philips Research, Moscow, Russia

Abstract. MRI scans appearance significantly depends on scanning protocols and, consequently, the data-collection institution. These variations between clinical sites result in dramatic drops of CNN segmentation quality on unseen domains. Many of the recently proposed MRI domain adaptation methods operate with the last CNN layers to suppress domain shift. At the same time, the core manifestation of MRI variability is a considerable diversity of image intensities. We hypothesize that these differences can be eliminated by modifying the first layers rather than the last ones. To validate this simple idea, we conducted a set of experiments with brain MRI scans from six domains. Our results demonstrate that 1) domain-shift may deteriorate the quality even for a simple brain extraction segmentation task (surface Dice Score drops from 0.85–0.89 even to 0.09); 2) fine-tuning of the first layers significantly outperforms fine-tuning of the last layers in almost all supervised domain adaptation setups. Moreover, fine-tuning of the first layers is a better strategy than fine-tuning of the whole network, if the amount of annotated data from the new domain is strictly limited.

Keywords: Domain adaptation · Segmentation · CNN · MRI

1 Introduction

Convolutional Neural Networks (CNN) are the most accurate segmentation methods for many medical image analysis tasks [18]. The core advantage of deep CNNs is their great flexibility due to a large number of trainable parameters. However, this flexibility may result in a dramatic drop in performance, if the test data comes from a different distribution which is a common situation for medical imaging. This fact is especially true for Magnetic Resonance Imaging (MRI) as different scanning protocols result in significant variations of slice orientation, thicknesses, and, most importantly, overall image intensities [10,16].

Electronic supplementary material The online version of this chapter (https://doi.org/10.1007/978-3-030-60548-3_12) contains supplementary material, which is available to authorized users.

Many of the existing MRI domain adaptation approaches rely on the information from the last CNN's layers, see details in Sect. 2. However, we assume that the differences in intensities can be successfully reduced by modifying the first convolutional layers. Surprisingly, this simple idea has not been directly compared with other fine-tuning strategies to the best of our knowledge. Therefore, we aimed to compare the following options in a series of supervised domain adaptation setups: fine-tuning the whole network, fine-tuning the first layers only and fine-tuning the last layers only. Our contribution is twofold:

- First, we show that publicly available dataset CC359 [19] holds a great potential for being utilised as a benchmark dataset for testing various DA approaches.
- Secondly, we prove that fine-tuning of the first layers outperforms significantly fine-tuning of the last layers. Moreover, fine-tuning of the first layers outperforms fine-tuning of the whole network when an extremely small amount of data is available.

Finally, we publish a complete experimental pipeline to provide a starting point for other researchers[1].

2   Related Work

A lot of domain adaptation methods exploit one of the following strategies:

1. Train a network on the source domain, then fine-tune it using data from the target domain [7,12,20,21].
2. Remove domain-specific information using an additional network head that aims to predict the scan domain. The core idea introduced in [6] is to minimize domain prediction accuracy rather than maximize by exploiting the gradient reversal layer (GRL). Related ideas were proposed in the medical image analysis community by [10,16].
3. Various ideas around Generative Adversarial Learning, e.g., [4,22].

Interestingly, at least some methods from all three groups exploit information from the first/last layers explicitly or implicitly and thus are connected to the core research question of our work.

The most widespread strategy is to fine-tune the last layers of the network. Though it's a natural solution for transfer learning as the generality of features tends to decrease with the number of the layer [23], several works showed promising results for domain adaptation [7,12,21]. In contrast, to the best of our knowledge, fine-tuning of the first layers was not properly researched for medical imaging. Moreover, the authors of [7,12] directly rejected the idea of fine-tuning the first layers because of an assumption that they pose too general, domain-independent characteristics. GRL-related algorithms usually minimize domain

---

[1] https://github.com/kechua/DART20.

shift by analyzing high-level features generated at the end of the network using similar motivation.

In the studies [5,13], dedicated to out-of-distribution detection (OOD), the best performance was achieved with the confidence scores computed for the intermediate or the last layers, but in those experiments out-of-distribution and in-distribution data came from the datasets of a very different nature (e.g. CIFAR-10 and SVHN), whereas in our case an "OOD sample" would correspond to a scan from a new domain.

Meanwhile, a study [1] of non-medical images showed that domain shift effects pop up at the very first layer. The authors train the net on the base domain and then assess domain shift by observing filter maps, produced by convolutions on a particular layer. The more pronounced domain shift on a certain layer is, the more domain-specific are the distributions of the filter maps on this layer. Concluding, that the first layers are susceptible to domain shift even more than the other layers, the authors then develop an unsupervised DA method, targeting the first layer only.

An idea somewhat similar to fine-tuning of the first layers was proposed in [11] for the setup of test-time domain adaptation. The key element of the pipeline, which is fine-tuned during the test time, is a shallow image-to-normalized-image CNN, which may be thought of as the first layers of a net, combined from the preprocessing and the main task nets.

Finally, the authors of [4] hypothesized that cross-modality (MRI-CT) domain shift causes significant changes mainly in the first layers, and developed an unsupervised domain adaptation framework based on adversarial learning. However, this idea wasn't validated directly in the paper.

3 Experiments

3.1 Data

We conduct all the experiments on a publicly available dataset CC359 [19]. It is composed of 359 MRIs of head acquired on scanners from three vendors (Siemens, Philips and General Electric) at both $1.5T$ and $3T$. Different combinations of a vendor and a field strength correspond to *six domains*. Data is equally distributed across domains, with an exception of Philips 1.5T domain, where only 59 subjects are present.

We do not apply any specific preprocessing to the data, except for two simple steps. First, we transform all the scans to the equal voxel resolution of $1 \times 1 \times 1$ mm via interpolation of the slices. Secondly, we scale the resulting images to the intensities of voxels between 0 and 1 before passing them to the network.

3.2 Metric

The quality of brain segmentation is usually measured with Dice Score [2].

$$\text{Dice Score} = \frac{2 \cdot |A \cap B|}{|A| + |B|}. \tag{1}$$

It measures voxel-wise similarity between two masks A and B:, which means that it captures volumetric quality of segmentation. However, in case of brain segmentation the most eloquent indicator of the model quality is how good the edges of the brain are segmented. Meanwhile, the edge zones account for a small share of the brain volume, which makes dice score not sensitive enough to the delineation quality. Therefore we use the Surface Dice metric [15], which compares how closely the segmentation and the ground truth surfaces align.

Surface Dice shares the same formula with Dice Score (Eq. 1), but A and B correspond to two surfaces in this case. The intersection of two surfaces depends on the *tolerance* value, which defines the maximum distance between predicted surface voxel and the ground truth surface voxel to be considered as matching elements. We report the experimental results with the tolerance of 1mm, since we find it sensitive enough to the changes in predictions of different methods.

3.3   Architecture and Training

Unlike 3D CNN architectures, 2D architectures give us an opportunity to investigate the behaviour of different approaches on an extremely small amount of labeled data from the target domain, i.e. on a subset of slices from one scan (see Sect. 3.4), thus we use 2D U-Net [17] in all our experiments. Besides, we have also conducted the baseline experiments for 3D U-Net [3] and observed equally pronounced decline in the segmentation quality.

U-Net [17] is one of the most widely used architectures which was originally developed for 2D image segmentation tasks. We adopt the original 2D U-Net model for our task introducing minor changes to keep up with the state-of-the-art level of architectures. We use residual blocks [8] instead of simple convolutions, for this was shown to improve segmentation quality [14]. We also apply convolutional layer with $1 \times 1 \times 1$ kernel to the skip-connections. We change the channel-wise concatenation at the end of the skip-connections to the channel-wise summation – it reduces memory consumption and preserves the number of channels in the following residual blocks. Our architecture is detailed in Fig. 1. We keep it without changes for the rest of the experiments. To reduce the dependence of the results on the architecture we also carried out the experiments for vanilla U-Net.

On the source domain, we train the model for 100 epochs, starting with the learning rate of 10^{-2} and reducing it to 10^{-3} at the epoch 80. When we transfer the model to the other domain we fine-tune it for 20 epochs, starting with the learning rate of 10^{-3} and reducing it to 10^{-4} at the epoch 15. Each epoch consists of 100 iterations of stochastic gradient descent with Nesterov momentum (0.9). At every iteration we sample a random slice and crop it randomly to the size of 256×256. Then we form a mini-batch of size 32 and pass it to the network. Training for 100 epochs takes about 4 h on a 16 GB nVidia Tesla V100 GPU. These are the GPUs installed on the Zhores supercomputer recently launched in Skolkovo Institute of Science and Technology (Skoltech) [24].

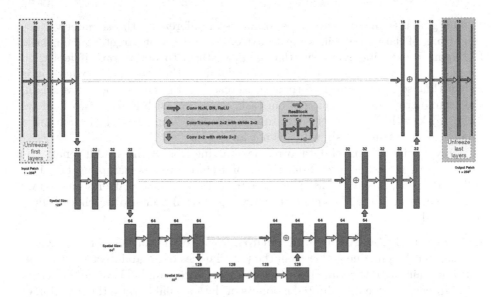

Fig. 1. The architecture of 2D U-Net [17] with minor modification we use in our work. In the scenarios implying freezing, either the first 3 or the last 3 convolutional layers are unfreezed. These layers contain an equal amount of filters (16) of the same size, which means that in both scenarios an equal number of parameters is fine-tuned.

3.4 Experimental Setup

Below we detail three main groups of the experiments we have carried out. By the term *scan* we always refer to the whole 3D MRI study, while *slice* is a 2D section of a *scan*.

The Baseline and the Oracle. First of all, we have to determine whether CC359 holds a potential for being useful for DA experiments. Thus we measure cross-domain model transferability on this data set. To do so, we train separately six models within the corresponding domains, and then test each model on the other domains. This forms the *baseline* of our study.

The test score of a model on the source domain is obtained via 3-fold cross validation. We refer to the result as *oracle*. It marks the upper boundary for all transferring methods; note, though, that in some cases transferring methods may outperform the oracle.

In the subsequent transferring experiments the models being transferred are trained on the whole source domain. We calculate the fraction of the gap between the oracle and the baseline that the method closes, because we find this way of measuring a method performance the most interpretable (discussed in detail in Sect. 4).

Supervised DA. In the main part of our study we consider three supervised domain adaptation strategies: fine-tuning of the whole model (*all layers*), fine-tuning of the *first layers* and fine-tuning of the *last layers*.

Our goal is to investigate the performance of different methods under various conditions of data availability, up to extreme shortage of target domain data. Preliminary experiments shows that quality starts to deteriorate if less than 5 scans are provided, which aligns perfectly with the study [21], where it is shown that segmentation performance decreases with the number of voxels in the ground truth mask. In case of our task, operating with a 2D network allows us to work with a subset of slices from a random scan instead of choosing a smaller ground truth mask, which is not an option.

We vary the amount of data available, starting from 3 and 1 additional MRI scans and then, making use of our choice of architecture, subsampling 1/2, 1/3, 1/6, 1/12, 1/24, 1/36 or 1/48 axial slices from 1 scan. In the latter scenarios we sample slices evenly with a constant step, e.g. in 1/3 scenario we choose slices 0, 3, 6 and so on.

Discussion of the Additional Setups. Despite the authors of [9] suggest focusing on the pipeline rather than the peculiarities of an architecture, we support our claim with the same line of experiments with vanilla U-Net architecture [17]. Moreover, extremely limited amounts of data available raise the question of augmentation, thus we repeat all the experiments for both architectures, introducing simple augmentation techniques: rotations and symmetric flips. We place all the results for vanilla U-Net and the results for the original net trained with augmentation in Supplementary Materials while discussing them in Sect. 4.

In our preliminary experiments we also tried other supervised DA setups. First, instead of fine-tuning, we trained the model from scratch on joint data from the source and the target domains. Secondly, we trained the model from scratch on data from the target domain only. We do not include aforementioned strategies in the further analysis for they yield extremely poor results.

Table 1. Cross-domain model transfer without fine-tuning. Column names are the source domains which the model is trained on, the row name is the target domains which the model is tested on. Sm, GE, Ph correspond to vendors, i.e. Siemens, GE and Philips. Results are given in surface Dice Score and the corresponding standard deviations are placed in the brackets.

	Sm, 1.5T	Sm, 3T	GE, 1.5T	GE, 3T	Ph, 1.5T	Ph, 3T
Sm, 1.5T	**.85 (.12)**	.51 (.15)	.72 (.08)	.56 (.13)	.71 (.10)	.71 (.07)
Sm, 3T	.72 (.08)	**.88 (.03)**	.70 (.07)	.67 (.10)	.63 (.10)	.66 (.06)
GE, 1.5T	.39 (.14)	*.09 (.05)*	**.87 (.05)**	*.30 (.10)*	.55 (.19)	.48 (.08)
GE, 3T	.80 (.05)	.63 (.13)	.66 (.10)	**.89 (.03)**	.67 (.10)	.67 (.06)
Ph, 1.5T	.63 (.08)	*.25 (.07)*	**.87 (.03)**	.43 (.06)	**.89 (.03)**	.46 (.08)
Ph, 3T	.54 (.13)	.34 (.13)	.70 (.11)	.37 (.10)	.47 (.14)	**.86 (.04)**

4   Results and Discussion

We show the presence of domain shift problem in Table 1. Evaluating 2D U–Net model on the source domain via cross-validation yields high Surface Dice values (diagonal elements or the *oracle*). Transferring the model without fine-tuning (non-diagonal elements or the *baseline*) leads to considerable quality deterioration. We emphasize the best scores with bold font and the worst with italics.

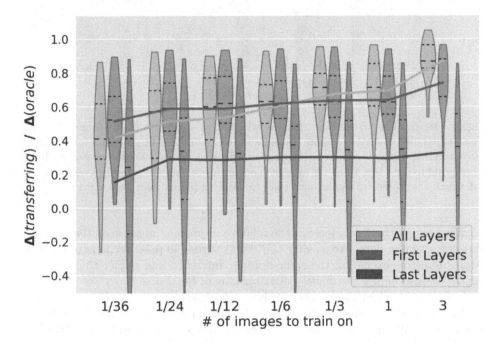

Fig. 2. Dependence of the relative Surface Dice improvement (y-axis) on the target domain data availability (x-axis) for the three transferring strategies. The lines correspond to the average scores. We also include distribution densities on 30 source-target pairs for every strategy.

To assess each DA method for a particular source-target pair we calculate the share of the *gap* between the oracle and the baseline this method closes. We denote this share D_R and define it the following way:

$$D_R = \frac{D - D_B}{D_O - D_B} = \frac{\Delta(transferring)}{\Delta(oracle)}, \tag{2}$$

where D_O is the oracle Surface Dice score on the target domain, D_B is the baseline score on the target domain and D is the score of a method being considered.

Below we compare three chosen approaches to supervised DA problem: fine-tuning of the whole model (*all layers*), fine-tuning of the *first layers* and fine-tuning of the *last layers*.

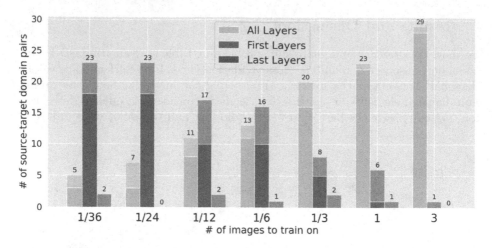

Fig. 3. Dependence of the best domain adaptation strategy on the target domain data availability. Each bar counts the number of the source-target pairs for which a corresponding method appears to be the most effective. The shadowed parts of the bars correspond to the cases, when the approach outperforms the others by the average score but below the statistical significance level (p-value of paired Sign Test < 0.1).

We consider the dependence of the relative improvement score on the amount of target training data. We average the scores across 30 possible pairs of source-target domains, excluding the same-domain inference and depict the trend in Fig. 2. We also depict the density distributions of the scores.

In Fig. 3 we report the number of the source-target pairs on which a selected method outperforms all the other methods (sums up to 30 over all methods for each set-up). We use paired sign test for every source-target pair to calculate the significance level. The instances for the sign test are the relative improvements of the Surface Dice scores different DA approaches yield for every single test image of the target domain.

Contrary to a mainstream conception, we show that fine-tuning of the first layers outperforms considerably fine-tuning of the last layers in our task. We therefore argue, that low-level features corresponding to the image intensity profile could be re-learned more efficiently than high-level features, which correspond to different brain structures and distinctive shapes.

Aside from freezing strategies comparison, we may see that under scarce data condition fine-tuning of the first layers becomes superior to fine-tuning of the whole model. It makes the former approach preferable in a highly practical setup, corresponding to the lack of annotated data in the target domain.

Substituting U-Net with residual blocks with vanilla U-Net or adding augmentation to either of them does not change the trends described. The results for those setups may be found in Supplementary materials.

5 Conclusion

We show a drastic reduction in segmentation quality for a naive model transfer between the domains of $CC359$. We hypothesize that the low-level feature maps of this data set are more prone to domain shift than the feature maps of deeper layers; hence the first layers are the primary source of the performance degradation.

We show that to be true by comparing different approaches to fine-tuning the network: fine-tuning the first layers outperforms fine-tuning the last layers. We also find that under the lack of annotated data for the target domain, fine-tuning of the first layers is superior to fine-tuning of the whole network.

Though we investigate a simple supervised setup, our results may suggest that unsupervised approaches will also benefit from targeting the first layers rather than the last ones.

References

1. Aljundi, R., Tuytelaars, T.: Lightweight unsupervised domain adaptation by convolutional filter reconstruction. In: Hua, G., Jégou, H. (eds.) ECCV 2016. LNCS, vol. 9915, pp. 508–515. Springer, Cham (2016). https://doi.org/10.1007/978-3-319-49409-8_43
2. Bakas, S., et al.: Identifying the best machine learning algorithms for brain tumor segmentation, progression assessment, and overall survival prediction in the brats challenge. arXiv preprint arXiv:1811.02629 (2018)
3. Çiçek, Ö., Abdulkadir, A., Lienkamp, S.S., Brox, T., Ronneberger, O.: 3D U-Net: learning dense volumetric segmentation from sparse annotation. In: Ourselin, S., Joskowicz, L., Sabuncu, M.R., Unal, G., Wells, W. (eds.) MICCAI 2016. LNCS, vol. 9901, pp. 424–432. Springer, Cham (2016). https://doi.org/10.1007/978-3-319-46723-8_49
4. Dou, Q., Ouyang, C., Chen, C., Chen, H., Heng, P.A.: Unsupervised cross-modality domain adaptation of convnets for biomedical image segmentations with adversarial loss. In: IJCAI (2018)
5. Erdil, E., Chaitanya, K., Konukoglu, E.: Unsupervised out-of-distribution detection using kernel density estimation. arXiv abs/2006.10712 (2020)
6. Ganin, Y., Lempitsky, V.: Unsupervised domain adaptation by backpropagation. In: International Conference on Machine Learning, pp. 1180–1189 (2015)
7. Ghafoorian, M., et al.: Transfer learning for domain adaptation in MRI: application in brain Lesion segmentation. In: Descoteaux, M., Maier-Hein, L., Franz, A., Jannin, P., Collins, D.L., Duchesne, S. (eds.) MICCAI 2017. LNCS, vol. 10435, pp. 516–524. Springer, Cham (2017). https://doi.org/10.1007/978-3-319-66179-7_59
8. He, K., Zhang, X., Ren, S., Sun, J.: Deep residual learning for image recognition. In: Proceedings of the IEEE Conference on Computer Vision and Pattern Recognition, pp. 770–778 (2016)
9. Isensee, F., Kickingereder, P., Wick, W., Bendszus, M., Maier-Hein, K.H.: No new-net. In: Crimi, A., Bakas, S., Kuijf, H., Keyvan, F., Reyes, M., van Walsum, T. (eds.) BrainLes 2018. LNCS, vol. 11384, pp. 234–244. Springer, Cham (2019). https://doi.org/10.1007/978-3-030-11726-9_21

10. Kamnitsas, K., et al.: Unsupervised domain adaptation in brain Lesion segmentation with adversarial networks. In: Niethammer, M., et al. (eds.) IPMI 2017. LNCS, vol. 10265, pp. 597–609. Springer, Cham (2017). https://doi.org/10.1007/978-3-319-59050-9_47

11. Karani, N., Chaitanya, K., Konukoglu, E.: Test-time adaptable neural networks for robust medical image segmentation. arXiv abs/2004.04668 (2020)

12. Kushibar, K., et al.: Supervised domain adaptation for automatic sub-cortical brain structure segmentation with minimal user interaction. Sci. Rep. **9**(1), 1–15 (2019)

13. Lee, K., Lee, K., Lee, H., Shin, J.: A simple unified framework for detecting out-of-distribution samples and adversarial attacks. In: NeurIPS (2018)

14. Milletari, F., Navab, N., Ahmadi, S.A.: V-Net: fully convolutional neural networks for volumetric medical image segmentation. In: 2016 Fourth International Conference on 3D Vision (3DV), pp. 565–571. IEEE (2016)

15. Nikolov, S., et al.: Deep learning to achieve clinically applicable segmentation of head and neck anatomy for radiotherapy. arXiv preprint arXiv:1809.04430 (2018)

16. Orbes-Arteaga, M., et al.: Multi-domain adaptation in brain MRI through paired consistency and adversarial learning. In: Wang, Q., et al. (eds.) DART/MIL3ID -2019. LNCS, vol. 11795, pp. 54–62. Springer, Cham (2019). https://doi.org/10.1007/978-3-030-33391-1_7

17. Ronneberger, O., Fischer, P., Brox, T.: U-Net: convolutional networks for biomedical image segmentation. In: Navab, N., Hornegger, J., Wells, W.M., Frangi, A.F. (eds.) MICCAI 2015. LNCS, vol. 9351, pp. 234–241. Springer, Cham (2015). https://doi.org/10.1007/978-3-319-24574-4_28

18. Shen, D., Wu, G., Suk, H.I.: Deep learning in medical image analysis. Annu. Rev. Biomed. Eng. **19**, 221–248 (2017)

19. Souza, R., et al.: An open, multi-vendor, multi-field-strength brain MR dataset and analysis of publicly available skull stripping methods agreement. NeuroImage **170**, 482–494 (2018)

20. Valindria, V.V., et al.: Domain adaptation for MRI organ segmentation using reverse classification accuracy. arXiv preprint arXiv:1806.00363 (2018)

21. Valverde, S., et al.: One-shot domain adaptation in multiple sclerosis lesion segmentation using convolutional neural networks. NeuroImage: Clin. **21**, 101638 (2019)

22. Yang, J., Dvornek, N.C., Zhang, F., Chapiro, J., Lin, M.D., Duncan, J.S.: Unsupervised domain adaptation via disentangled representations: application to cross-modality liver segmentation. In: Shen, D., et al. (eds.) MICCAI 2019. LNCS, vol. 11765, pp. 255–263. Springer, Cham (2019). https://doi.org/10.1007/978-3-030-32245-8_29

23. Yosinski, J., Clune, J., Bengio, Y., Lipson, H.: How transferable are features in deep neural networks? In: Advances in Neural Information Processing Systems, pp. 3320–3328 (2014)

24. Zacharov, I., et al.: 'Zhores' - petaflops supercomputer for data-driven modeling, machine learning and artificial intelligence installed in Skolkovo Institute of Science and Technology. Open Eng. **9**, 512–520 (2019)

DCL 2020

Siloed Federated Learning for Multi-centric Histopathology Datasets

Mathieu Andreux$^{(\boxtimes)}$, Jean Ogier du Terrail, Constance Beguier,
and Eric W. Tramel

Owkin, Inc., New York, NY, USA
{mathieu.andreux,jean.du-terrail,constance.beguier,
eric.tramel}@owkin.com

Abstract. While federated learning is a promising approach for training deep learning models over distributed sensitive datasets, it presents new challenges for machine learning, especially when applied in the medical domain where multi-centric data heterogeneity is common. Building on previous domain adaptation works, this paper proposes a novel federated learning approach for deep learning architectures via the introduction of local-statistic batch normalization (BN) layers, resulting in collaboratively-trained, yet center-specific models. This strategy improves robustness to data heterogeneity while also reducing the potential for information leaks by not sharing the center-specific layer activation statistics. We benchmark the proposed method on the classification of tumorous histopathology image patches extracted from the Camelyon16 and Camelyon17 datasets. We show that our approach compares favorably to previous state-of-the-art methods, especially for transfer learning across datasets.

Keywords: Federated learning · Histopathology · Heterogeneity · Deep learning

1 Introduction

Federated learning (FL) has recently emerged as a new paradigm for scalable and practical privacy-preserving machine learning (ML) on decentralized datasets [22]. In the case of medical data, notably digital histopathology images, this approach brings the promise of ML architectures trained over large and

---

M. Andreux and J. O. du Terrail contributed equally to this work.

Electronic supplementary material The online version of this chapter (https://doi.org/10.1007/978-3-030-60548-3_13) contains supplementary material, which is available to authorized users.

diverse populations, a necessary component for truly generalizable medical findings. By bridging the gap between localized, curated, and non-portable per-institution datasets, a federated approach permits the study of otherwise unconstructible research datasets. With the ability to utilize large and rich datasets for ML, medical researchers have the potential to make new scientific discoveries, as evidenced in [8].

Two challenges currently limit the applicability of existing FL techniques to real-world histopathology datasets: inter-center data heterogeneity and well-understood privacy assurances on communicated model parameters. Statistical heterogeneity in the data distribution between participating centers is a key issue for FL [14,18,24], which may lead to model biases or even prevent training convergence [24]. Such heterogeneity is often present in histopathology datasets, where variations in staining procedure, scanning device configuration, and systematic imaging artifacts are commonplace [16].

In this paper, we propose a novel federated learning strategy, called *SiloBN*, which brings improvements both in terms of resilience to heterogeneity and privacy. This strategy relies on batch normalization (BN) [13] layers, which are ubiquitous in deep learning (DL), especially for computer vision applications. *SiloBN* introduces, or uses already specified, *local-statistic* BN layers within DL architectures. More precisely, we propose the following novel contributions:

i) We demonstrate the applicability of FL to real-world tile-level digital pathology image classification, using the Camelyon16 and Camelyon17 datasets [21] (Sect. 5). To the best of our knowledge, this work is among the first examples thereof. Previous works rather relied on the SplitLearning framework [30].

ii) We introduce *SiloBN*, a new FL approach for training DL models robust to inter-center data variability by introducing local-statistic BN layers. This approach also reveals less sensitive information by not communicating local activation statistics (Sect. 4).

iii) We show that our proposed approach achieves same or better performance than existing federated techniques for intra-center generalization (Sect. 5.2), even in challenging settings, and yields better out-of-domain generalization results (Sect. 5.3).

2   Background

In the years since the publication of the 2013 ICPR-winning approach of [7], the application of DL to digital pathology tasks has blossomed [9,30]. In parallel, federated learning [22,26], a distributed privacy-preserving ML paradigm, has recently experienced incredible growth as a topic of study, garnering much interest in the medical research community. Due to the highly sensitive nature of the medical data, techniques such as FL are a requirement for investigators in order to develop state-of-the-art ML models over a fractured and highly regulated data landscape.

Federated Learning. As presented in [22], the aim of FL is to obtain a single well-trained model from a distributed network of K participants, each possessing their own privately-held datasets. Generically, this is accomplished by optimizing model parameters θ w.r.t. an expectation over the individual participant losses ℓ_i, i.e. $\mathcal{L}(\theta) \triangleq \mathbb{E}_i [\ell_i(\theta)]$. This optimization is carried out iteratively in federated rounds, where at each round t the expectation is approximated as $\sum_{i \in \mathcal{C}^{(t)}} \alpha_i \ell_i(\theta)$, where $\mathcal{C}^{(t)} \sim [K]$ is random sub-sampling of participants and $\alpha_i \geq 0$ are contribution weightings. The authors propose *federated averaging* (*FedAvg*) to reduce coordination rounds by performing several local optimization steps prior to aggregation,

$$\overbrace{\theta_i^{(t)} = \mathrm{Opt}_E \left(\theta^{(t)}, \ell_i\right),}^{\text{Local}} \qquad \overbrace{\theta^{(t+1)} = \sum_{i \in \mathcal{C}^{(t)}} \frac{N_i}{N^{(t)}} \theta_i^{(t)},}^{\text{Central Server,}} \qquad (1)$$

where N_i is the number of data samples at participant i, and $N^{(t)} \triangleq \sum_{i \in \mathcal{C}^{(t)}} N_i$, $\mathrm{Opt}_E(\cdot)$ is an E-step iterative optimization, and t indexes federated rounds. For brevity, we refer the reader to the original work of [22] for details. Further adaptations can be made to the algorithm to enhance privacy, e.g. [1,4], but are out of scope of the present work.

Data Heterogeneity in FL. Data heterogeneity has been identified as a key open challenge for FL [18]. For example, despite its practical success, *FedAvg* does not provide a natural guarantee of convergence, especially for highly dissimilar participant datasets. The *FedProx* algorithm of [24], similar to the *EASGD* of [31], introduces a quadratic loss term at each round $\ell_i^{(t)}(\theta_i) = \ell_i(\theta_i) + \frac{\lambda}{2}||\theta_i - \theta^{(t)}||^2$, to restrain local models from diverging during local optimization. This approach was shown to be effective on the large-K, heterogeneous LEAF datasets [6]. Novel participant sampling strategies have also been recently proposed [10,19] to help adapt *FedAvg* to heterogeneous participant datasets. Finally, one can also seek to partition the set of participants into clusters of effective collaborators, as in [25]. It should be noted that the assumption that data samples are independent and identically distributed (*i.i.d.*) is core to many ML techniques, and empirical risk minimization [29] in particular. For FL, we retain an *i.i.d.* assumption on data *intra*-participant, but not *inter*-participant.

3   FL in Healthcare

The original application of FL sought to define techniques for distributed training over mobile edge devices, e.g. smart phones [17,22], and has now been practically scaled to $K > 10^6$ [5]. Notably, this application also assumes that each participant has a paucity of training data, i.e. $N_i \ll N = \sum_{j=1}^K N_j \; \forall i \in [K]$. The application of FL to the context of distributed and private *medical datasets* reverses the order of magnitude on both of these variables, with the number of participants often being quite low, e.g. $K \leq 10$, and generally having access to much larger and higher quality datasets at each center, i.e. $\log N_i \approx \log N$. The implication of this reversal is that the nature and goals of the federated system shift between the two settings:

i) *Consumer FL*—Federated training coordinated and of value to central service provider. No single participant, out of very many, can train an effective model, and no participant has an incentive to use such a local model. Goal of the service provider is to coordinate and distribute final global model back to participants. Service provider may hold their own private or public evaluation data.

ii) *Collaborative FL*—Federated training coordinated by central service provider *on behalf* of few participants, who value and seek to reuse its outputs. Participants have enough local data of quality to produce effective local models, but seek to augment this local task performance through collaboration. Evaluation data is sensitive and held by participants.

For this work, we consider the *collaborative FL setting* for training a tile-level classifier for WSI across multiple centers. This setting poses several challenges for the construction of distributed algorithms which are not fully addressed in the literature described in Sect. 2.

Data heterogeneity is a key challenge for collaborative FL, as inconsistencies in data preparation will not be averaged out over a large population of participants. Additionally, in the consumer FL setting, $|\mathcal{C}(t)| \ll K$, which means outlier participants are rarely revisited. For collaborative FL, such strong sub-sampling is not desirable, as per-round updates would become *more* biased and also lead to potential privacy challenges as the sensitivity of the aggregated model w.r.t. the center of origin increases. For this reason, we consider only full-participation federated rounds in the case of collaborative FL. In order to keep stochasticity in the optimization, contrary to [22] which uses local epochs, we use E to denote the number of local *mini-batch* updates in (1).

4   Proposed Method

Batch Normalization (BN). BN [13] is a critical component of modern deep learning architectures. A BN layer renormalizes an input tensor x as

$$\mathrm{BN}(x) = \gamma \frac{x - \mu}{\sqrt{\sigma^2 + \epsilon}} + \beta, \tag{2}$$

where μ and σ^2, hereafter denoted as BN statistics, are calculated as the running means and variances, respectively, of each channel computed across both spatial and batch dimensions, and γ and β are *learned* affine renormalization parameters, and where all computations are performed along the channel axis. BN has been used to both speed model training and also enhance model predictive performance [13].

While BN layers are common architectural components [11,12,28], they have not been thoroughly addressed in the federated setting, and are often simply ignored or removed [22]. Indeed, the naïve application of *FedAvg* would ignore the different roles of the local activation statistics (μ, σ^2), and the trained renormalization (γ, β) and simply aggregate both at each federated round, as depicted in Fig. 1 (left). In the following, we use this method as a *baseline* when using *FedAvg* on a network with BN layers.

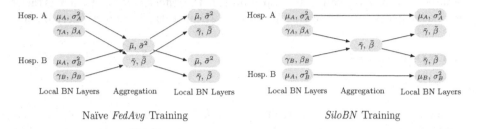

Naïve *FedAvg* Training   *SiloBN* Training

Fig. 1. Description of the different approaches to multi-center training of BN layers for two hospitals. In this description, all variables follow the definitions given in (2). Computation flows from left to right. Non-BN layers are shared in both methods.

BN layers can also be used to separate local and domain-invariant information. Specifically, the BN statistics and learned BN parameters play different roles [20]: while the former encode *local* domain information, the latter can be transferred across domains. This opens interesting possibilities to tune models to local datasets, which we discuss in the following section.

SiloBN. Instead of treating all BN parameters equally, we propose to take into account the separate roles of BN statistics (μ, σ^2) and learned parameters (γ, β). Our method, called *SiloBN*, consists in only sharing the learned BN parameters across different centers, while *BN statistics remain local*. Parameters of non-BN layers are shared in the standard fashion. This method is depicted in Fig. 1 (right). Keeping BN statistics local permits the federated training of a model robust to the heterogeneity of the different centers, as the *local statistics ensure that the intermediate activations are centered to a similar value across centers*. While *SiloBN* can be applied to models which already possess BN layers, we demonstrate that in some cases, BN layers can be added to the base model to improve resilience to heterogeneity as well as domain generalization.

Model Personalization and Transfer. A consequence of the proposed method is that one model per center is learned instead of a single global model, thereby enabling model personalization. However, it is not straightforward to generalize the resulting models to unseen centers, as BN statistics must be tuned to the target dataset. A simple approach to overcome this issue is to follow the *AdaBN* [20] method, and recompute BN statistics on a data batch of the new target domain, while all other model parameters remain frozen to those resulting from the federated training.

Privacy and BN. When applying standard FL methods on a network containing BN layers, the BN statistics are shared between centers and may leak sensitive information from local training datasets. With *SiloBN*, the BN statistics are not shared among centers, and in particular the central server never has access to all the network parameters. As a number of privacy attacks [27,32] are most effective with white-box, full-parameter, knowledge, the reduction of the amount of shared information can only help diminish the effectiveness of such attacks when compared to synchronizing batch activation statistics across participants.

5   Experiments

5.1   Datasets

In order to understand the performance of federated algorithms in a practical setting, we seek real-world datasets with actual multi-centric origins. For histopathology, the Camelyon 16 [3] and Camelyon 17 [2] challenge datasets provide H&E stained whole slide images (WSI) of lymph node sections drawn from breast cancer patients from 2 and 5 hospitals, respectively. Figure 4 in Supp. Mat. shows representative slides from each center, which have different color statistics on average.

Building on Camelyon 16 and Camelyon 17, we construct two tile-level tumor classification histopathology datasets, referred to as FL-C16 and FL-C17. In both cases, the task consists in classifying the tiles between tumorous and healthy ones. In order to get tile-level annotations, tumorous tiles are extracted from pixel-level annotated tumorous WSI, whenever available, while healthy tiles are extracted from healthy WSI. This restriction, motivated by incomplete expert annotations, ensures tile-level labels' correctness. For each WSI, we extract at most 10,000 non-overlapping fixed-size matter-bearing tiles uniformly at random using a U-Net segmentation model [23] trained for matter detection. To reduce class imbalance, for each healthy WSI, we cap the number of extracted tumor-negative tiles to 1,000 for FL-C16 and 100 for FL-C17, as FL-C17 has fewer annotated tumor-positive tiles than FL-C16. Finally, each dataset is partitioned into a training set (60%), a validation set (20%), and a test set (20%) using per-hospital stratification. For FL-C16 (resp. FL-C17), tiles from same slide (resp. tiles from same patient) are put into the same partition. Table 3 in Supp. Mat. details the distribution of tiles in the data partitions.

5.2   Experiments on FL Histopathology Datasets

Baselines. We compare the proposed *SiloBN* method to two standard Federated algorithms, *FedAvg* and *FedProx*, which treat BN statistics as standard parameters. Additionally, we report results obtained in the *pooled* setting, where the training sets of each center are concatenated and the origin center information is discarded during training. To measure the benefits of the personalization of our model, we also compare the proposed method to a *local* training, where on each center one model is trained independently without inter-center collaboration.

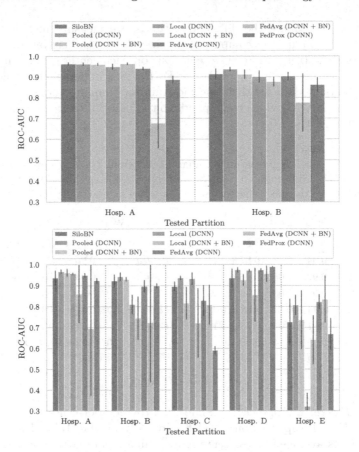

Fig. 2. Predictive performance of different training approaches used for training DCNN with and without BN layers. Performance is measured on per-center held-out test sets. Note that the *Pooled*, *FedProx*, and *FedAvg* approaches produce a single model which is evaluated on multiple test sets, while the *Local* and *SiloBN* approaches produce one model per center, which is then evaluated on its corresponding center-specific test set.

Training Setting. For tile classification, we use two deep convolutional neural network (DCNN) architectures that only differ by the presence or absence of BN layers, which we denote as DCNN and DCNN+BN, respectively. These architectures are detailed in Fig. 3 in Supp. Mat. Note that, due to computational constraints, this architecture is shallower than standard ones [11,12,28], yielding better performance without BN layers, which is not entirely realistic. Each FL algorithm is run with $E = 1$ or $E = 10$ local batch updates. Each training session is repeated 5 times using different random seeds for weight initialization and data ordering. Finally, all local optimization is performed using Adam [15]. Table 4 in Supp. Mat. provides all related training hyperparameters.

Evaluation. To compare the different methods, we measure intra-center generalization performance by computing the area under the curve (AUC) of the ROC curve of the trained model's predictions on each center's held-out data for single-model methods (*Pooled*, *FedAvg*, or *FedProx*), and personalized models are tested on held-out data for their specific training domain (*Local* and *SiloBN*). We also report mAUC, which corresponds to the pan-center mean of intra-center generalization AUCs.

Table 1. Mean AUC over federated centers for different training approaches on FL-C16 and FL-C17. The proposed *SiloBN* is on par or better than other FL methods, and is shown to be the most effective approach when using BN layers.

		WITH BN LAYERS				WITHOUT BN LAYERS	
	E	FL-C16	FL-C17		E	FL-C16	FL-C17
Pooled	–	0.94 ± 0.03	0.87 ± 0.11	*Pooled*	–	0.95 ± 0.02	0.93 ± 0.07
Local	–	0.92 ± 0.05	0.76 ± 0.16	*Local*	–	0.93 ± 0.03	0.80 ± 0.25
SiloBN	1	0.94 ± 0.03	0.86 ± 0.13				
	10	$\mathbf{0.94 \pm 0.03}$	$\mathbf{0.88 \pm 0.10}$	*FedProx*	10	0.87 ± 0.03	0.81 ± 0.16
FedAvg	1	0.81 ± 0.05	0.70 ± 0.18	**FedAvg**	1	0.62 ± 0.15	0.80 ± 0.17
	10	0.73 ± 0.14	0.80 ± 0.22		**10**	$\mathbf{0.92 \pm 0.02}$	$\mathbf{0.89 \pm 0.07}$

Impact of Federated Approaches. Figure 2 provides per-center testing performance for $E = 10$ updates on FL-C16 and FL-C17 and Table 1 shows a full table of results averaged across centers, including a comparison of performance over the number of local steps per federated round. On both datasets, the proposed *SiloBN* outperforms other FL strategies for the DCNN+BN network, and is on par with the pooled and local trainings. For the DCNN network, for which *FedAvg* and *SiloBN* are equivalent, we can make the same observation. However, the lack of BN layers increases the sensitivity of *FedAvg* to E. We note that for FL-C17, some reported results in Fig. 2 have very large error bars or very low AUC. More analysis is required to understand the source of these instabilities. Since some of these odd results occur for local trainings, they could e.g. stem from poor local data quality. Overall, from this set of experiments, we conclude that the proposed *SiloBN* is the only method that can fully utilize BN layers in the FL setting.

5.3   Out-of-Domain Generalization Experiments

We now investigate the ability of the proposed method to transfer to new domains. We first train a model on FL-C16 and test it on FL-C17. We compare the proposed *SiloBN*, where testing is done with *AdaBN* as explained in Sect. 4, to a DCNN without BN layers trained with *FedAvg*, for which no adaptation is needed. Per-hospital generalization results are provided in Table 2. We note that the proposed algorithm outperforms the *FedAvg*-trained DCNN model on average across centers. Moreover, we note that its generalization performance is much more stable, both across trainings (lower variance) and across centers.

Table 2. Domain generalization of *SiloBN* and *FedAvg* (DCNN) from FL-C16 to FL-C17. On average, *SiloBN* outperforms *FedAvg* and yields more stable results.

Hospital	A	B	C	D	E	*Mean*
SiloBN+AdaBN	0.94 ± 0.01	$\mathbf{0.91 \pm 0.04}$	0.93 ± 0.01	$\mathbf{0.98 \pm 0.01}$	$\mathbf{0.95 \pm 0.01}$	0.94 ± 0.02
FedAvg (DCNN)	$\mathbf{0.95 \pm 0.02}$	0.86 ± 0.03	$\mathbf{0.95 \pm 0.01}$	$\mathbf{0.98 \pm 0.02}$	0.88 ± 0.03	0.92 ± 0.05

6   Conclusion

We have introduced *SiloBN*, a novel FL strategy which relies on *local-statistic* BN layers in DCNNs. Experiments on real-world multicentric histopathology datasets have shown that this method yields similar, or better, intra-center generalization capabilities than existing FL methods. Importantly, the proposed approach outperforms existing methods on out-of-domain generalization in terms of performance and stability, while enjoying a better privacy profile by not sharing local activation statistics. Future works could quantify the privacy benefits of this approach as well as study its applicability to other domains, e.g. radiology.

References

1. Abadi, M., et al.: Deep learning with differential privacy. In: Proceedings ACM Conference on Computer and Communications Security, October 2016
2. Bandi, P., et al.: From detection of individual metastases to classification of lymph node status at the patient level: the camelyon17 challenge. IEEE Trans. Med. Imaging **38**(2), 550–560 (2018)
3. Bejnordi, B.E., et al.: Diagnostic assessment of deep learning algorithms for detection of lymph node metastases in women with breast cancer. JAMA **318**(22), 2199–2210 (2017)
4. Bonawitz, K., et al.: Practical secure aggregation for privacy-preserving machine learning, iACR Cryptology Preprint (2017)
5. Bonawitz, K., et al.: Proceedings SysML Conference, Palo Alto, CA (2019)
6. Caldas, S., et al.: Leaf: a benchmark for federated settings. arXiv preprint arXiv:1812.01097 (2018)
7. Cireşan, D.C., Giusti, A., Gambardella, L.M., Schmidhuber, J.: Mitosis detection in breast cancer histology images with deep neural networks. In: Mori, K., Sakuma, I., Sato, Y., Barillot, C., Navab, N. (eds.) MICCAI 2013. LNCS, vol. 8150, pp. 411–418. Springer, Heidelberg (2013). https://doi.org/10.1007/978-3-642-40763-5_51
8. Courtiol, P., et al.: Deep learning-based classification of mesothelioma improves prediction of patient outcome. Nat. Med. **25**(10), 1519–1525 (2019)

9. Dimitriou, N., Arandjelović, O., Caie, P.D.: Deep learning for whole slide image analysis: an overview. Front. Med. **6** (2019)
10. Goetz, J., Malik, K., Bui, D., Moon, S., Liu, H., Kumar, A.: Active federated learning (2019). arXiv Preprint [cs.LG]:1909.12641
11. He, K., Zhang, X., Ren, S., Sun, J.: Deep residual learning for image recognition. In: The IEEE Conference on Computer Vision and Pattern Recognition (CVPR), pp. 770–778, June 2016
12. Hu, J., Shen, L., Sun, G.: Squeeze-and-excitation networks. In: Proceedings of the IEEE Conference on Computer Vision and Pattern Recognition, pp. 7132–7141 (2018)
13. Ioffe, S., Szegedy, C.: Batch normalization: accelerating deep network training by reducing internal covariate shift. arXiv preprint arXiv:1502.03167 (2015)
14. Kairouz, P., et al.: Advances and open problems in federated learning (2019). arXiv Preprint [cs.LG]:1912.04977
15. Kingma, D., Ba, J.: Adam: a method for stochastic optimization, December 2014. arXiv Preprint [cs.LG]:1412.698
16. Komura, D., Ishikawa, S.: Machine learning methods for histopathological image analysis. Comput. Struct. Biotechnol. J. **16**, 34–42 (2018)
17. Konecný, J., McMahan, H.B., Ramage, D., Richtárik, P.: Federated optimization: distributed machine learning for on-device intelligence, October 2016. arXiv Preprint [cs.LG]:1610.02527
18. Li, T., Sahhu, A.K., Talwalkar, A., Smith, V.: Federated learning: challenges, methods, and future directions (2019). arXiv Preprint [cs.LG]:1908.07873
19. Li, X., Huang, K., Yang, W., Wang, S., Zhang, Z.: On the convergence of FedAvg on non-IID data (2019). arXiv Preprint [stat.ML]:1907.02189
20. Li, Y., Wang, N., Shi, J., Liu, J., Hou, X.: Revisiting batch normalization for practical domain adaptation. arXiv preprint arXiv:1603.04779 (2016)
21. Litjens, G., et al.: 1399 H&E-stained sentinel lymph node sections of breast cancer pateints: the CAMELYON dataset. GigaScience **7**(6), giy065 (2018)
22. McMahan, H.B., Moore, E., Ramage, D., Hampson, S., Arcas, B.A.: Communication-efficient learning of deep networks from decentralized data, February 2017. arXiv Preprint [cs.LG]:1602.05629
23. Ronneberger, O., Fischer, P., Brox, T.: U-net: convolutional networks for biomedical image segmentation. In: Navab, N., Hornegger, J., Wells, W.M., Frangi, A.F. (eds.) MICCAI 2015. LNCS, vol. 9351, pp. 234–241. Springer, Cham (2015). https://doi.org/10.1007/978-3-319-24574-4_28
24. Sahu, A.K., Li, T., Sanjabi, M., Zaheer, M., Talwalkar, A., Smith, V.: On the convergence of federated optimization in heterogeneous networks (2018). arXiv Preprint [cs.LG]: 1812.06127
25. Sattler, F., Müller, K.R., Samek, W.: Clustered federated learnig: model-agnostic distributed multi-task optimization under privacy constraints (2019). arXiv Preprint [cs.LG]:1910.01991
26. Shokri, R., Shmatikov, V.: Privacy-preserving deep learning. In: Proceedings of the 22nd ACM SIGSAC Conference on Computer and Communications Security, pp. 1310–1321. ACM (2015)
27. Shokri, R., Stronati, M., Song, C., Shmatikov, V.: Membership inference attacks against machine learning models. In: 2017 IEEE Symposium on Security and Privacy (SP), pp. 3–18. IEEE (2017)

28. Tan, M., Le, Q.: EfficientNet: rethinking model scaling for convolutional neural networks. In: Proceedings of the 36th International Conference on Machine Learning. Proceedings of Machine Learning Research, vol. 97, pp. 6105–6114. PMLR, Long Beach, 09–15 June 2019. http://proceedings.mlr.press/v97/tan19a.html
29. Vapnik, V.: Principles of risk minimization for learning theory. In: Advances in Neural Information Processing Systems, pp. 831–838 (1992)
30. Vepakomma, P., Gupta, O., Dubey, A., Raskar, R.: Reducing leakage in distributed deep learning for sensitive health data. In: AI for Social Good ICLR Workshop, May 2019
31. Zhang, S., Choromanska, A.E., LeCun, Y.: Deep learning with elastic averaging SGD. In: Advances in Neural Information Processing Systems, pp. 685–693 (2015)
32. Zhu, L., Liu, Z., Han, S.: Deep leakage from gradients. In: Advances in Neural Information Processing Systems, December 2019

On the Fairness of Privacy-Preserving Representations in Medical Applications

Mhd Hasan Sarhan[1,2](✉)(iD), Nassir Navab[1,3], Abouzar Eslami[2](iD),
and Shadi Albarqouni[1,4](iD)

[1] Computer Aided Medical Procedures, Technical University of Munich,
Munich, Germany
hasan.sarhan@tum.de
[2] Translational Research Lab, Carl Zeiss Meditec AG, Munich, Germany
[3] Computer Aided Medical Procedures, Johns Hopkins University, Baltimore, USA
[4] Computer Vision Lab, ETH Zurich, Zürich, Switzerland

Abstract. Representation learning is an important part of any machine learning model. Learning privacy-preserving discriminative representations that are invariant against nuisance factors is an open question. This is done by removing sensitive information from the learned representation. Such privacy-preserving representations are believed to be beneficial to some medical and federated learning applications. In this paper, a framework for learning invariant fair representations by decomposing the learned representation into target and sensitive codes is proposed. An entropy maximization constraint is imposed on the target code to be invariant to sensitive information. The proposed model is evaluated on three applications derived from two medical datasets for autism detection and healthcare insurance. We compare with two methods and achieve state of the art performance in sensitive information leakage trade-off. A discussion regarding the difficulties of applying fair representation learning to medical data and when it is desirable is presented.

Keywords: Fair representations · Invariant representations · Privacy-preserving representations · ABIDE dataset

1 Introduction

With the success of machine learning algorithms in learning descriptive representations, it became necessary that such representations are robust against arbitrary noisy signals. Such representations are defined to hold meaningful information about a target task of interest while being invariant to attributes believed to be noisy or sensitive. Deep neural networks have shown tremendous performance in learning descriptive representations. Since these networks learn from human collected data, they might suffer from learning inherent biases and leak information that is not intended to be released. Such systems are becoming

S. Albarqouni et al. (Eds.): DART 2020/DCL 2020, LNCS 12444, pp. 140–149, 2020.
https://doi.org/10.1007/978-3-030-60548-3_14

corner stones of various decision support systems including medical ones. This has raised various concerns about fairness, privacy and discrimination in medical applications and statistical models [2, 10]. Privacy-preserving representations and federated learning present a possible solution [14]. In most literature [15, 16, 18], a fair representation is treated as a representation z that is well informed about a target attribute y (good bank credit) while being robust against a nuisance sensitive attribute s (age or gender). The assumption when building such models is that sensitive s and target y variables are independent. This assumption does not always hold and it is highly application specific. When s and y are not independent, which is the case in most fairness applications, there is a trade-off between target accuracy and sensitive information leakage (such as salary prediction with gender as a sensitive attribute) [15, 18]. In medical applications, it is crucial to distinguish these two cases as such fairness might hinder the performance of the downstream task. If a diagnostic decision is dependent on the subject's race, it might be unfair to build a diagnostic model that ignores or hides such information. We show in our experiments an example of both cases.

To achieve fairness, earlier approaches relied on data-massaging techniques [5, 13]. Such approaches make the assumption that equal opportunities in training data would generalize to test data. The data points are rearranged to enforce a fair representation of all groups. Such approaches are difficult to ensure especially in medical applications where data is scarce and must be utilized. Later, a fair clustering approach is proposed to learn a representation where examples are clustered to have equal number of examples from all sensitive attributes in each cluster [19]. This approach hinders the representations distribution property with clustering constraints which leads to a larger target/sensitive trade-off. Later approaches rely on approximating mutual information between representation z and sensitive attribute s and minimizing it in an either adversarial [9, 15, 18] or non-adversarial [7, 11, 16] manner. To approach invariance non-adversarially, Louizos et al. [7] have proposed the variational fair autoencoder (VFAE) where a representation is learned to be representative of the data while robust to a nuisance. This representation is then used for a target task. The robustness is approached by Maximum Mean Discrepancy (MMD) penalty. In [11], the authors use a variational upper bound to approximate mutual information between z and s. Other approaches, tackle the problem in adversarial manner by posing a zero-sum minimax game between an encoder that learns the desired representation and an adversarial discriminator that learns to distinguish the sensitive information in the representation [18]. Such zero-sum optimization problems might need additional regularization terms to stabilize the training [15]. A non-zero sum game approach is proposed to mitigate this problem by optimizing an entropy loss that enforces the discriminator to be agnostic to sensitive information [15]. While adversarial approaches have been successful in learning fair representations, they suffer from convergence relate problems and sometimes they could hinder the performance [11].

In this work, we propose a non-adversarial framework for learning fair representations by imposing an entropy maximization constraint on a decomposed latent representation that encodes target information in one split and residual information related sensitive attributes in the other. The representations are learned in a multi-task-like framework with entropy loss to enforce the target representation to be agnostic to sensitive information. We evaluate this framework on two medical datasets and three applications related to fairness. We discuss how such approaches could be applied to medical data and how fairness approaches could be beneficial or harmful in the healthcare.

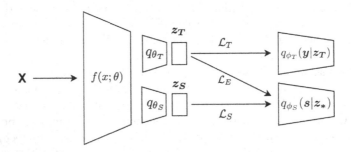

Fig. 1. The proposed fairness framework. The input is split into target and residual sensitive representations and then fed into two discriminators to predict the target and sensitive attributes.

2   Method

let $\mathcal{X}$ be a dataset of subjects and $x \in \mathbb{R}^D$ be an input sample. An input sample x is associated with a target attribute (class) $y = \{y_1, \ldots, y_n\} \in \mathbb{R}^n$, and a sensitive class $s = \{s_1, \ldots, s_m\}$. We aim to learn a representation z_T that contains information about the target attributes while limiting the information leakage about sensitive attributes. An encoder is trained to map the input x to two representations $z_T \in \mathbb{R}^d$, $z_S \in \mathbb{R}^d$ where z_T contains target information while z_S carry residual information partly related to the sensitive attributes.

An input sample x is given to a shared encoder $f(x; \theta)$, then projected into two representations z_T and z_S (see Fig. 1). These subspaces are acquired by the two encoders $q_{\theta_T}(z_T|x)$, and $q_{\theta_S}(z_S|x)$ parameterized by θ_T and θ_S respectively. The first part of the proposed loss is analogous to multi-task problem loss. Each of the two representations is fed to a network to learn a task. The target code z_T is fed to a target discriminator $q_{\phi_T}(y|z_T)$ to learn the target downstream task. The residual sensitive code z_S is fed to a sensitive discriminator $q_{\phi_S}(s|z_S)$ to learn to distinguish sensitive attributes. All these networks are trained in a supervised fashion to minimize the following objective functions,

$$\mathcal{L}_T(\theta, \theta_T, \phi_T) = -\sum p(y|x) \log[q_{\phi_T}(y|z_T)], \tag{1}$$

$$\mathcal{L}_S(\theta_S, \phi_S) = -\sum p(s|x) \log[q_{\phi_S}(s|z_S)], \tag{2}$$

where $p(\boldsymbol{y}|\boldsymbol{x})$ is the target attribute $\boldsymbol{y}$ of input $\boldsymbol{x}$, $p(\boldsymbol{s}|\boldsymbol{x})$ is the sensitive attribute $\boldsymbol{s}$ of input $\boldsymbol{x}$, and θ, θ_T, θ_S, ϕ_T, ϕ_S are the network parameters (weights). Only the parameters of the sensitive discriminator (ϕ_S) and the sensitive encoder (θ_S) are optimized when minimizing $\mathcal{L}_S$. The parameters of the backbone shared encoder are not considered. Since we assume independence of the two attributes (target and sensitive), the losses $\mathcal{L}_S$ and $\mathcal{L}_T$ in this case have competing objectives.

This on its own, would not prevent leakage of sensitive information to $\boldsymbol{z}_T$. Inspired by adversarial entropy maximization proposed by Roy *et al.* [15], we maximize the entropy of the sensitive discriminator given the target representation as input. In the proposed framework, it is not required to maximize entropy in an adversarial manner since the target and sensitive codes are split and the entropy maximization could be applied only to the target representation as

$$\mathcal{L}_E(\theta, \theta_T, \phi_S) = \sum q_{\phi_S}(\boldsymbol{s}|\boldsymbol{z}_T) \log[q_{\phi_S}(\boldsymbol{s}|\boldsymbol{z}_T)], \qquad (3)$$

The entropy loss $\mathcal{L}_E$ enforces the target representation to be uninformative of the sensitive task. This is a desired feature of a fair representation since we favour representations that mitigate sensitive information leakage.

The overall objective function of the proposed framework can be written as

$$arg \min_{\theta, \theta_T, \theta_S, \phi_T, \phi_S} \mathcal{L}_T(\theta, \theta_T, \phi_T) + \mathcal{L}_S(\theta_S, \phi_S) + \lambda_E \mathcal{L}_E(\theta, \theta_T, \phi_S) \qquad (4)$$

where λ_E is a hyper-parameter to weigh the *Entropy* loss.

3   Experiments

In this section, we evaluate the effectiveness of the learned fair representations in medical applications by evaluating on two publicly available datasets for three tasks. The results are compared with a baseline model and two state-of-the-art models in learning fair representations in an adversarial fashion. We also discuss the quantitative results of the learned representations by showing the t-SNE projections of the learned representations compared to the baseline.

3.1   Experimental Setup

Datasets: To evaluate fair classification in medical applications we chose two publicly available datasets to investigate. First, we use the tabular *Heritage Health* dataset[1]. The dataset contains physicians records and insurance claims of over 60k patients. The target downstream task is to predict a comorbidity indicator that estimates the risk of patient death in the next several years (the Charlson Index). The sensitive attribute in this case is the sex of the patient, partially similar to [17]. We study in this case if sex should be overlooked while analyzing insurance claims of re-visitations of patients to the hospitals.

---

[1] Heritage Health dataset, www.heritagehealthprize.com.

The second dataset we use is the *ABIDE* dataset [4]. The dataset contains fMRI scans and demographic data of 1112 subjects. Similar to [1,12], we utilize the same set of 871 subjects that met the quality criteria. The dataset contains 403 subjects with Autism Spectrum Disorder (ASD) and 468 healthy ones. We follow the prepossessing steps as in [1], namely the Configurable Pipeline for the Analysis of Connectomes (C-PAC) [3]. The target downstream task is the detection of ASD subjects. We experiment with two different potential sensitive attributes namely sex and age. We binarize age into young and old subjects. We consider patients over 20 years old to be in one class.

Implementation Details: For the health heritage dataset, we use an encoder with three hidden layers (64, 64, 40). The target classifier and discriminator are networks with three hidden layers (10, 10, 10). The size of the representation is 20. The learning rate for all components is 10^{-5} and weight decay is 5×10^{-4}. The dataset is split into train\validation splits (80%\20%) and all models are trained on the same split for fair comparison. $\lambda_E = 1.5$.

For the ABIDE dataset, we use an encoder with two hidden layers (3000, 1500, 200) with dropout probability of 0.3. The target and discriminator are 2-hidden-layered networks (40, 30). The learning rate for all components is 10^{-5} and eight decay is 5×10^{-4}. The baseline models for both datasets are the encoder and the target models combined to solve one task which is target attribute identification. Hence, no residual sensitive representation z_S is learned in this scenario. Following [12], we use a 10-fold stratified cross-validation for evaluation. All models are trained on the same splits for fair comparison. $\lambda_E = 2$. We use Adam optimizer [6] for all models. Moreover, we use the same encoder and discriminators architectures for training the state-of-the-art methods with the same hyper-parameters. We set $\alpha = 0.3$ for the compared methods [15,18].

Experiments Design: In the experiments we aim to evaluate the learned representations in terms of 1) how much sensitive information is retained in the target representation z_T. To this end, we use the learned representation z_T as an input to as simple logistic regression model trained to identify sensitive attributes from target representations. Ideally, this model is not able to distinguish sensitive attributes and has an accuracy similar to naively guessing the majority class as output. It is also important to evaluate 2) how well the learned representation z_T describes the target attributes. We train a logistic regression model to identify target attributes from the target representation z_T. Ideally, this model should perform on par or better than a baseline model.

3.2   ABIDE Dataset

For the ABIDE dataset we chose two sensitive attributes for the experiments. First we use age as a sensitive attribute. It is believed that age is a redundant phenotypic variable that does not contribute positively to the prediction task of ASD [12]. This goes inline with the definition of fair representations where the sensitive information is believed to be independent from the target task and could

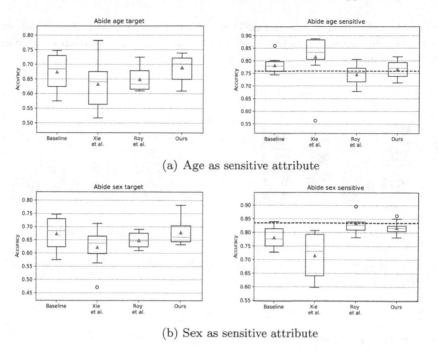

(a) Age as sensitive attribute

(b) Sex as sensitive attribute

Fig. 2. Results on the ABIDE dataset. The majority classifier is represented by a dashed black line.

act as noise signal that disturbs the target classification. We try with this experiment to remove any information about patients' age from the pre-processed imaging input by learning a representation that is free from this information. The phenotypic age information itself is not provided to the network as part of the input. The results for age are shown in Fig. 2a. Regarding the sensitive information leakage mitigation, the proposed approach well hides the sensitive information in the representation. This can be seen by the mean accuracy on the sensitive class (76.57%) which is the closest to the majority classifier (75.88%). The mean and median target accuracy (69%, 70.11%) are well improved over the baseline (67.38%, 68.4%) and close to the recent state-of-the-art mean accuracy on this dataset [12] (70%) using population graph convolutional neural networks which includes phenotypic information into their model. As suggested by [12] and validated in our experiment, not all phenotypic information are beneficial to the models and some might have negative effect. By removing the age information from the data representation, we obtain better classification for the ASD detection target task. We do not see a similar trend considering sex as the sensitive attribute Fig. 2b. When hiding sex related information from the representation, we monitor a good performance in mitigating sensitive information leakage. However, the target accuracy does not have a similar trend and it is even lowered in median value (66.09%) compared to the baseline (68.4%).

This could indicate that sex and ASD are not independent variables which suggests that there should be a trade-off between sensitive information leakage and target performance.

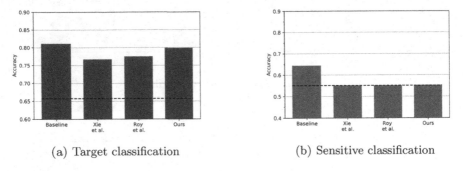

(a) Target classification   (b) Sensitive classification

Fig. 3. Results on the health dataset. The majority classifier is represented by a dashed black line.

3.3   Health Dataset

The results on the health dataset are shown in Fig. 3. The proposed model better preserves the sensitive information compared to the baseline classifier, where the baseline has an accuracy of 64.38% compared to 55.22% which is closer to the majority classifier line at 55.09%. The results of the proposed model are similar to other fair models in terms of hiding sensitive information. However, the proposed model has a better trade-off for the target accuracy. This is shown by the higher target accuracy of the model compared to other fairness models while keeping on par sensitive performance. This suggests that it is possible to make an informed decision if a patient will return to the hospital or not regardless of the subject's sex.

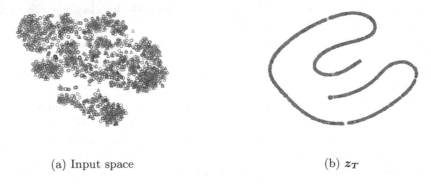

(a) Input space   (b) z_T

Fig. 4. t-SNE projections of the input space and z_T on the health dataset

Qualitative Analysis: The learned embeddings are visualized with t-SNE [8] projections on the health dataset in Fig. 4 (Better seen in high resolution). We visualize the input space and z_T projections to intuitively understand what information is held in each representation. The color represents the sensitive information (sex) and the shape represents the target information (Charlson Index). We can see in the input space there is no separation between target information as the shapes are mixed. The sensitive information is indistinguishable to some extent as we observe some small clusters with more dominance of one color. When projecting the learned embeddings, it is clear that all sensitive information is mixed. The target information is split as could be seen at the two ends of the projections where each end is dominated by one shape.

4   Discussion and Conclusion

In this paper, a model for learning privacy-preserving invariant representations is proposed. The model learns invariance by decomposing representations into target and residual representations and enforcing entropy maximization constraints to mitigate sensitive information leakage. The model is evaluated on two healthcare-related datasets to answer the question of the effectiveness of using fair representations in medical applications. One dataset is the tabular insurance claims health dataset, and the other is the ABIDE dataset for detecting autism spectrum disorder from fMRI images. We distinguish different cases for using fairness in medical applications, these cases are mainly relying on our belief of fairness in these situations and how much the selected sensitive attributes contribute to diagnosis decision. In the health dataset, the target task is not a diagnostic decision but rather, an expectation of the subject's hospital readmission. This information is important for insurance companies for the subjects' future plans. In such a scenario, some historic biases might be included in the retrospectively collected data. Such biases could be learned by a machine learning model and the model would have similar biases cause by target and sensitive attributes that are not independent. The proposed fairness model mitigates the leakage of sensitive information at the cost of a little drop in target performance. This suggests that the two variables cannot be forced to be independent (at least by the proposed and compared models). On the other hand, the ABIDE dataset is a diagnostic dataset which changes the way fairness is approached. It is important in such cases for the selected sensitive information to be irrelevant for the diagnosis (i.e. target and sensitive attributes are independent). The age at the scan time is selected as a sensitive attribute since it is reported in the literature that including this phenotypic information in the classification drops the diagnostic performance of the model [12]. This suggests that such an attribute is irrelevant to the diagnosis and might act as a nuisance. We validate this by showing that target performance improves when removing this attribute that is believed to be clinically irrelevant to the case at hand. However, this is not the case when the attribute is clinically relevant to the diagnosis. When trying to hide information about the subject's sex, the target performance drops which could be because the two variables are not independent. These experiments show that using privacy-preserving models for medical applications should be done with clinical relevance in mind as hiding clinically important information for the sake of security might compromise the downstream task performance.

Acknowledgments. S.A. is supported by the PRIME programme of the German Academic Exchange Service (DAAD) with funds from the German Federal Ministry of Education and Research (BMBF).

References

1. Abraham, A., et al.: Deriving reproducible biomarkers from multi-site resting-state data: an autism-based example. NeuroImage **147**, 736–745 (2017)
2. Chen, I.Y., Szolovits, P., Ghassemi, M.: Can AI help reduce disparities in general medical and mental health care? AMA J. Ethics **21**(2), 167–179 (2019)
3. Craddock, C., et al.: Towards automated analysis of connectomes: the configurable pipeline for the analysis of connectomes (C-PAC). Front. Neuroinform. **42** (2013)
4. Di Martino, A., et al.: The autism brain imaging data exchange: towards a large-scale evaluation of the intrinsic brain architecture in autism. Mol. Psychiatry **19**(6), 659–667 (2014)
5. Kamiran, F., Calders, T.: Classifying without discriminating. In: 2009 2nd International Conference on Computer, Control and Communication, pp. 1–6. IEEE (2009)
6. Kingma, D.P., Ba, J.: Adam: a method for stochastic optimization. arXiv preprint arXiv:1412.6980 (2014)
7. Louizos, C., Swersky, K., Li, Y., Welling, M., Zemel, R.: The variational fair autoencoder. arXiv preprint arXiv:1511.00830 (2015)
8. Maaten, L.V.D., Hinton, G.: Visualizing data using t-SNE. J. Mach. Learn. Res. **9**, 2579–2605 (2008)
9. Madras, D., Creager, E., Pitassi, T., Zemel, R.: Learning adversarially fair and transferable representations. arXiv preprint arXiv:1802.06309 (2018)
10. Mehrabi, N., Morstatter, F., Saxena, N., Lerman, K., Galstyan, A.: A survey on bias and fairness in machine learning. arXiv preprint arXiv:1908.09635 (2019)
11. Moyer, D., Gao, S., Brekelmans, R., Galstyan, A., Ver Steeg, G.: Invariant representations without adversarial training. In: Advances in Neural Information Processing Systems, pp. 9084–9093 (2018)
12. Parisot, S., et al.: Disease prediction using graph convolutional networks: application to autism spectrum disorder and Alzheimer's disease. Med. Image Anal. **48**, 117–130 (2018)
13. Pedreshi, D., Ruggieri, S., Turini, F.: Discrimination-aware data mining. In: Proceedings of the 14th ACM SIGKDD International Conference on Knowledge Discovery and Data Mining, pp. 560–568 (2008)
14. Rieke, N., et al.: The future of digital health with federated learning. arXiv preprint arXiv:2003.08119 (2020)
15. Roy, P.C., Boddeti, V.N.: Mitigating information leakage in image representations: a maximum entropy approach. In: Proceedings of the IEEE Conference on Computer Vision and Pattern Recognition, pp. 2586–2594 (2019)
16. Sarhan, M.H., Navab, N., Eslami, A., Albarqouni, S.: Fairness by learning orthogonal disentangled representations. arXiv preprint arXiv:2003.05707 (2020)

17. Song, J., Kalluri, P., Grover, A., Zhao, S., Ermon, S.: Learning controllable fair representations. arXiv preprint arXiv:1812.04218 (2018)
18. Xie, Q., Dai, Z., Du, Y., Hovy, E., Neubig, G.: Controllable invariance through adversarial feature learning. In: Advances in Neural Information Processing Systems, pp. 585–596 (2017)
19. Zemel, R., Wu, Y., Swersky, K., Pitassi, T., Dwork, C.: Learning fair representations. In: International Conference on Machine Learning, pp. 325–333 (2013)

Inverse Distance Aggregation
for Federated Learning
with Non-IID Data

Yousef Yeganeh[1]([⊠]), Azade Farshad[1], Nassir Navab[1,2], and Shadi Albarqouni[1,3]

[1] Computer Aided Medical Procedures, Technical University of Munich,
Munich, Germany
y.yeganeh@tum.de
[2] Whiting School of Engineering, Johns Hopkins University, Baltimore, USA
[3] Department of Computing, Imperial College London, London, UK

Abstract. Federated learning (*FL*) has been a promising approach in the field of medical imaging in recent years. A critical problem in *FL*, specifically in medical scenarios is to have a more accurate shared model which is robust to noisy and out-of distribution clients. In this work, we tackle the problem of statistical heterogeneity in data for *FL* which is highly plausible in medical data where for example the data comes from different sites with different scanner settings. We propose **IDA** (**I**nverse **D**istance **A**ggregation), a novel adaptive weighting approach for clients based on meta-information which handles unbalanced and non-iid data. We extensively analyze and evaluate our method against the well-known *FL* approach, Federated Averaging as a baseline.

Keywords: Deep learning · Federated learning · Distributed learning · Privacy-preserving. · Heterogeneous data · Robustness

1 Introduction

Federated learning (*FL*) was proposed as a decentralized learning scheme where the data in each client is private and not exposed to other participants, yet they contribute to generation of a shared (global) model in a server that represents the clients' data [12]. An aggregation strategy in the server is essential in *FL* for combining the models of all clients. Federated Averaging (*FedAvg*) [21] is one of the most well-known *FL* methods which uses the normalized number of samples in each client to aggregate the models in the server. Another aggregation approach using temporal weighting along with a synchronous learning strategy was proposed in [3]. Many recent approaches have been proposed in order to improve the generalization or personalization of the global model using the ideas of knowledge transfer, knowledge distillation, multi-task learning and meta-learning [1, 2, 4, 8, 9, 15, 29].

Project page: https://ida-fl.github.io/.

S. Albarqouni et al. (Eds.): DART 2020/DCL 2020, LNCS 12444, pp. 150–159, 2020.
https://doi.org/10.1007/978-3-030-60548-3_15

Even though *FL* has emerged into a promising and popular method to engage with privacy preserving distributed learning, it has faced some challenges: **a)** Expensive communication, **b)** privacy, **c)** systems heterogeneity and **d)** statistical heterogeneity[16]. Although a large number of recent works on *FL* such as [20,25] are focused on communication efficiency due to its application on edge devices with unstable connections [16], commonly using approaches such as compressed networks or compact features, its most determining aspects in the medical field are data privacy and heterogeneity [11,23]. Data heterogeneity assumption includes: **a)** Massively distributed: The data points are distributed among a very large number of clients. **b)** Non-iid (Not independent and identically distributed): Data in each node comes from a distinct distribution. The local data points are not representative of the whole data distribution (combination of all clients' data). **c)** Unbalancedness: The number of samples across clients has a high variance. Such heterogeneity is foreseeable in medical data due to many reasons, for example, class imbalance in pathology, intra-/inter-scanner variability (domain shift), intra-/inter-observer variability (noisy annotations), multi-modal data, and different tasks for clients.

There has been numerous works to handle each of these data assumptions [10]. Training a global model with *FL* in non-iid data is a challenging task. Model training in deep neural network suffers quality loss and may even diverge given non-iid data [5]. There has been multiple works dealing with this problem. Sattler et al. [24] propose clustering loss terms and using cosine similarity to overcome the divergence problem when clients have different data distributions. Zhao et al. [33] overcome the non-iid problem by creating a subset of data which is shared globally with the clients. In order to maintain system heterogeneity (affected by their main idea of nonuniform local updates), FedProx [17] proposes a proximal term to minimize the distance between the local and global models. Close to our approach, geometric median is used in [22] to decrease the effect of corrupted gradients on the federated model.

In the last few years, there has been a growing interest in applying *FL* in healthcare, in particular, to medical imaging. Sheller et al. [27] were among the first works who applied *FL* to multi-institutional data for Brain Tumor Segmentation task. To date, there has been numerous works on *FL* in Healthcare [7,18,19,26,32]. However, little attention has been paid to the aggregation mechanism given the data and system heterogeneity; for example, when the data is non-iid, or the participation rate of the clients is pretty low.

In this work, we try to overcome the challenges of statistical heterogeneity in data and propose a robust aggregation method at the server side (*cf.* Fig. 1). Our weighting coefficients are based on the meta-information extracted from the statistical properties of the model parameters. Our goal is to train a low variance global model given high variance local models which is robust to non-iid and unbalanced data. Our contributions are twofolds; **a)** A novel adaptive weighting scheme for federated learning which is compatible with other aggregation approaches, **b)** Extensive evaluation of different scenarios on non-iid data on multiple datasets.

Next, a brief overview of the federated learning concept is introduced in the methodology section before diving into the main contribution of the paper, the Inverse Distance Aggregation (IDA). Experiments and results on both machine learning datasets (Proof-of-Concept), and clinical use-cases are demonstrated and discussed.

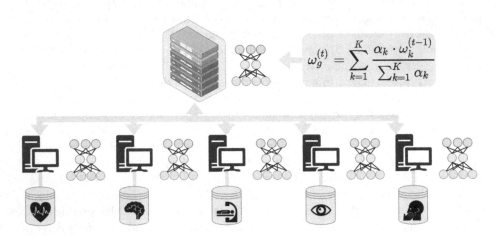

Fig. 1. Federated learning with non-iid data - the data has different distributions among clients.

2   Method

Given a set of K clients with their own data distribution $p_k(x)$ and a shared neural network with parameters ω, the objective is to train a global model minimizing the following objective function;

$$arg \min_{\omega_g^t} f(x; \omega_g^t), \quad \text{where} \quad f(x; \omega_g^t) = \sum_{k=1}^{K} f(x; \omega_k^t), \tag{1}$$

where ω_g^t, ω_k^t are the global and local parameters, respectively.

2.1   Client

Each randomly sampled client, from the total number of K clients (based on the participation rate pr), receives the global model parameter ω_g^t at communication round t, and trains the shared model, initialized by ω_g^t, on its own training data $p_k(x)$ for E iterations to minimize its local objective function $f_k(x) = \mathbf{E}_{x \sim p_k(x)}[f(x; \omega_k^t)]$ where ω_k^t is the weight parameters of the client k. The training data in each client is a subset of the whole training data, which can be sampled from different classes of data. The number of classes of data assigned to each client is denoted by n_{cc}.

2.2   Server

Each round t, the updated local parameters ω_k^t are sent back to the server and aggregated to form the updated global parameter ω_g^t,

$$\omega_g^t = \sum_{k=1}^{K} \alpha_k \cdot \omega_k^{t-1}. \tag{2}$$

where α_k is the weighting coefficient. This procedure continues for the given total communication rounds T.

2.3   Inverse Distance Aggregation (IDA)

In order to reduce the inconsistency among the updated local parameters due to the non-iid problem, we propose a novel robust aggregation method, denoted as Inverse Distance Aggregation (**IDA**). The core of our method is the way the coefficients α_k are computed, which is based on the inverse distance of each client parameters to the average model of all clients. This allows us to reject or weigh less the models who are poisoning, *i.e.* out-of-distribution models.

To realize this, the ℓ_1-norm is utilized as a metric to measure the distance of clients ω_k to the average one ω_{Avg} as

$$\alpha_k = \frac{1}{Z}\|\omega_{Avg}^{t-1} - \omega_k^{t-1}\|^{-1}, \tag{3}$$

where $Z = \sum_{k\in K}\|\omega_{Avg}^{t-1} - \omega_k^{t-1}\|^{-1}$ is a normalization factor. In practise, we add ϵ to both numerator and denominator to avoid any numerical instability. Note that $\alpha_k = 1$ when clients' parameters is equivalent to the average one, and $\alpha_k = n_k$ is equivalent to the *FedAvg* [21].

We also propose to use the training accuracy of clients in the final weighting which we denote by INTRAC (INverse TRaining ACcuracy) to penalize over-fitted models and encourage under-trained models in the aggregated model. To calculate the coefficients for INTRAC, We assign $\alpha'_k = \frac{Z'}{max(\frac{1}{K},acc_k)}$. The *max* function is used to assure all of the values are above chance level. Here acc_k is the training accuracy of client k, α'_k is the INTRAC coefficient and $Z' = \sum_{k\in K} max(\frac{1}{K},acc_k)$ is the normalization factor. We normalize the calculated coefficients α'_k once again to bring them to the range of $(0,1]$. To combine different coefficient values (i.e. INTRAC, IDA, FedAvg), we multiply the acquired coeffecients and normalize them in the range of $(0,1]$.

3   Experiments and Results

We evaluated our method on commonly used databases to show a Proof-of-Concept (PoC) before we present some results on a clinical use-case. We compare the results of our method **IDA** against the baseline method *FedAvg* [21]. In the first set of PoC experiments, we investigate the following: 1) Non-iid vs. iid: Comparison of *FedAvg* and **IDA** in iid and non-iid with different datasets and architectures. 2) Ablation study: Investigation of effectiveness of IDA compared to FedAvg 3) Sensitivity analysis: Performance comparison in extreme situations.

Datasets We show the results of our evaluation on cifar-10 [13], fashion-mnist (f-mnist) [31] and HAM10K(multi-source dermatoscopic images of pigmented lesions)[30] datasets. f-mnist is a well-known variation of mnist with $50k$ images of 28×28 black and white clothing pieces. cifar-10 is another dataset with $60k$ 32×32 images of vehicles and animals, commonly used in computer vision. For the clinical study, we evaluate our method on HAM10k dataset which includes a total number of 10015 images of different pigmented skin lesions in 7 classes. The different classes and their number of samples in HAM10k are as follows: Melanocytic nevi: 6705, Melanoma: 1113, Benign keratosis: 1099, Basal cell carcinoma: 514, Actinic Keratoses: 327, Vascular: 142, Dermatofibroma: 115. We chose this dataset due to its heavy unbalancedness.

Implementation Details. The training settings for each dataset are: LeNet [14] for f-mnist with 10 classes, batchsize=128, learning rate (lr) = 0.05 and local iteration of 1 (E = 1), VGG11 [28] without batch normalization and dropout layers for cifar-10 with 10 classes and batchsize = 128, lr = 0.05 and E = 1. For HAM10K, we used Densenet-121 [6] with 7 classes, batchsize = 32, lr = 0.016 and E = 1. In all of the experiments 90% ofr the images are randomly sampled for training and the rest are employed for evaluation. All of the models are trained for a total number of 5000 rounds. The mentioned values are the default for all experiments unless otherwise specified.

Evaluation Metrics. In all of the experiments, we separate a part of each client's dataset as its test set, and we report the accuracy of the global (aggregated) model on the union of the test sets of clients and the local accuracy of each client on it's own local test data. This gives us an indication of how well the global model is representative of the aggregated dataset. We report the classification accuracy in all of the experiments.

3.1   Proof-of-Concept

Non-iid vs. Iid. In this section we evaluate and compare **IDA** with *FedAvg* on f-mnist and cifar-10 datasets given different scenarios of data distribution in clients. Table 1 demonstrates the results of balanced data distribution where all clients have the same or similar number of samples for $n_{cc} \in \{3, 5, 10(iid)\}$ and $pr \in \{30\%, 50\%, 100\%\}$. Our results show that **IDA** has slightly better or on-par performance to *FedAvg* in all scenarios of balanced data distribution.

Ablation Study. In this section, we investigate the effect of different components of the weighting coefficients. We evaluate all of the proposed components on cifar-10 and f-mnist and compare them with two baseline methods, namely *FedAvg*, and another baseline where $\alpha_k = 1$, denoted by *Mean* shown in Table 2. We also evaluate the combination of our weighting method with number of samples per client (**IDA + FedAvg**) and adding the training accuracy of each client to the weighting scheme (**IDA + INTRAC**). The results indicate that combining different weighting schemes can lead to a better performing global model in

Table 1. Comparison between our method and the baseline on cifar10 and f-mnist with different number of classes per client in non-iid and iid scenarios

Dataset	n_{cc} / pr / Method	3c			5c			iid
		30%	50%	100%	30%	50%	100%	100%
cifar-10	FedAvg	63.20	65.11	69.81	19.68	83.11	80.94	87.77
	IDA	**64.36**	**67.70**	**70.80**	**76.06**	**83.55**	**83.82**	**89.46**
f-mnist	FedAvg	86.23	87.09	**87.45**	87.60	87.81	87.16	86.95
	IDA	**87.64**	**87.61**	87.44	**87.93**	**87.89**	**87.46**	**87.10**

FL. This supports our hypothesis, that if some of the clients have lower quality or poisonous models, *FedAvg* would be vulnerable, but our methods can lower the contribution of bad models (overfitted, low quality or poisonous models) so the final model performs better on the federated dataset.

Table 2. Ablation study on different weighting combinations on f-mnist and cifar-10 datasets.

Method	Settings		
	f-mnist | $n_{cc} = 3$ | $pr = 30\%$	cifar-10 | $n_{cc} = 3$ | $pr = 30\%$	
	K = 10	K = 10	K = 20
Mean	87.47	65.82	84.80
FedAvg	86.23	63.20	22.84
IDA	87.64	64.36	83.98
IDA + FedAvg	86.67	**67.29**	82.14
IDA +INTRAC	**88.33**	64.93	**85.23**

Sensitivity Analysis. In real-life scenarios, stability of learning process in unfavorable conditions is critical. In *FL* it is not mandatory for the members to contribute in each round, so the participation rate can be different in each round of training, and we might have lower quality models in any round. It is very likely that some clients have very few data samples, and some other clients have a lot of data. In this section we investigate the global model's performance given low participation rate and severe non-iidness.

Low Participation Rate in Non-iid Distribution. To investigate the effect of participation rate, we used 1000 clients on f-mnist dataset with (batchsize = 30, $lr = 0.016$ and $n_{cc} = 3$ and each client has up to 500 samples). In this experiment, we observe that despite the fact that this dataset is relatively easy to learn, decreasing the participation rate of clients lowers the performance (*cf.* Fig. 2). When the participation rate is at 1%, the model trained using *FedAvg* collapses. However, when we increase the participation rate to 5% the model continues to learn. We observe a robust performance for both **IDA** and **IDA + FedAvg** in both scenarios.

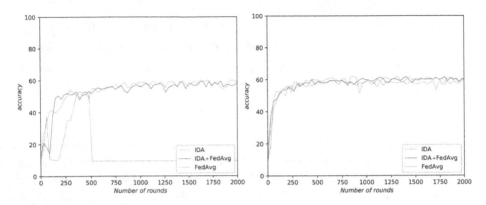

Fig. 2. Left: participation rate (pr) of 0.01; Right: participation rate of 0.05. The pr affects the stability of federated learning, and it is shown that **IDA** has stable performance comparing to FedAvg.

Severity of Non-IID. To analyze the effect of non-iidness on the performance of our method, we design an experiment by increasing the data samples of the low performing clients. To achieve this, first we train our models in a normal fashion as mentioned in previous sections. Then we choose three clients with the lowest accuracy at the end of the initial training and double the amount of their samples in the training data distribution. We repeat the training using the newly generated data distribution. We propose this experiment to see the effect of *FedAvg* weighting in a scenario where low performing clients are given higher weight. It can be seen in Fig. 3 that before increasing the number of samples, **IDA** performs marginally better compared to other methods; however, after we increase the number of samples in those three clients, *FedAvg* collapses at the beginning of training. Considering the performance of *Mean* aggregation, we see that **IDA** is the main contributing factor to the learning process.

3.2   Clinical Use-Case

We evaluate our proposed method on HAM10k dataset and show our results in Table 3. Even though the global accuracy of the model using **IDA** is on par with *FedAvg*, it can be seen that the local accuracy (accuracy of clients on their own test set) using **IDA** is superior to *FedAvg* in all scenarios. This indicates that **IDA** has a better generalization and lower variance in local accuracy of clients.

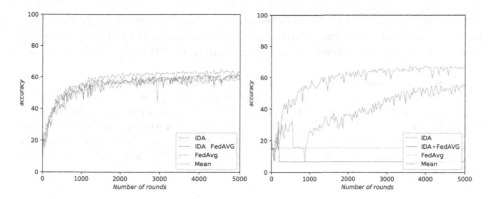

Fig. 3. Accuracy of global model of clients with non-iid data distribution on cifar-10: in the right we have the same clients, and the same learning hyperparameters of the left, but the number of samples in three of the clients with poor performances increased. The local distribution of data points in those three clients remained the same. This experiment is performed on cifar-10 dataset with $K = 10$ clients, $n_{cc} = 3$, $E = 2$, lr = 0.01 and random number of samples per class per client up to 1000 samples.

Table 3. Investigation on an unbalanced data distribution among the clients in federated setting, with five random classes per client, and random number of samples per client for HAM10k.

Method	n_{cc}	Global accuracy	Local accuracy
FedAvg	1	**69.72**	60.52 ± 9.20
IDA	1	69.16	**61.21 ± 8.79**
FedAvg	2	**62.23**	57.14 ± 10.84
IDA	2	61.21	**60.21 ± 5.48**
FedAvg	10 (iid)	63.5	52.88 ± 15.73
IDA	10 (iid)	**63.72**	**57.38 ± 10.56**

4   Discussion and Conclusion

In this work, we proposed a novel weighting scheme for aggregation of client models in a federated learning setting for non-iid and unbalanced data distribution. Our weighting is calculated based on the statistical meta-information which gives higher weights in aggregation to the clients that their data has a lower distance to the global average. We also propose another weighting approach called INTRAC that normalizes models to lower the contribution of overfitted models to the shared model. Our extensive experiments show that our proposed method outperforms FedAvg in terms of classification accuracy in non-iid scenario. Our proposed method is also resilient to low quality or poisonous data in the clients. For instance, if the majority of clients are rather aligned, then they can rule out the out-of-distribution models. This is not the case with FedAvg, however, which is based on the presumption that the clients with more data, have a better distribution compared to other models, and they should have more voting power in the global model. Future research directions concerning the out-of-distribution models detection and robust aggregation schemes should be further considered.

Acknowledgements. S.A. is supported by the PRIME programme of the German Academic Exchange Service (DAAD) with funds from the German Federal Ministry of Education and Research (BMBF). A.F. is supported by Munich Center for Machine Learning (MCML) with funding from the German Federal Ministry of Education and Research (BMBF) under Grant No. 01IS18036B. We also gratefully acknowledge the support of NVIDIA Corporation with the donation of the Titan V GPU used for this research.

References

1. Beel, J.: Federated meta-learning: Democratizing algorithm selection across disciplines and software libraries. Science (AICS) **210**, 219 (2018)
2. Chen, F., Dong, Z., Li, Z., He, X.: Federated meta-learning for recommendation. arXiv preprint arXiv:1802.07876 (2018)
3. Chen, Y., Sun, X., Jin, Y.: Communication-efficient federated deep learning with layerwise asynchronous model update and temporally weighted aggregation. IEEE Trans. Neural Netw. Learn. Syst. (2019)
4. Corinzia, L., Buhmann, J.M.: Variational federated multi-task learning. arXiv preprint arXiv:1906.06268 (2019)
5. Hsieh, K., Phanishayee, A., Mutlu, O., Gibbons, P.B.: The Non-IID data quagmire of decentralized machine learning. arXiv preprint arXiv:1910.00189 (2019)
6. Huang, G., Liu, Z., Van Der Maaten, L., Weinberger, K.Q.: Densely connected convolutional networks. In: Proceedings of the IEEE Conference on Computer Vision and Pattern Recognition, pp. 4700–4708 (2017)
7. Huang, L., et al.: Patient clustering improves efficiency of federated machine learning to predict mortality and hospital stay time using distributed electronic medical records. J. Biomed. Inform. **99**, 103291 (2019)
8. Jeong, E., Oh, S., Kim, H., Park, J., Bennis, M., Kim, S.L.: Communication-efficient on-device machine learning: Federated distillation and augmentation under Non-IID private data. arXiv preprint arXiv:1811.11479 (2018)
9. Jiang, Y., Konečný, J., Rush, K., Kannan, S.: Improving federated learning personalization via model agnostic meta learning. arXiv preprint arXiv:1909.12488 (2019)
10. Kairouz, P., et al.: Advances and open problems in federated learning. arXiv preprint arXiv:1912.04977 (2019)
11. Kaissis, G.A., Makowski, M.R., Rückert, D., Braren, R.F.: Secure, privacy-preserving and federated machine learning in medical imaging. Nature Mach. Intell., 1–7 (2020)
12. Konečný, J., McMahan, B., Ramage, D.: Federated optimization: distributed optimization beyond the datacenter. arXiv preprint arXiv:1511.03575 (2015)
13. Krizhevsky, A., Hinton, G., et al.: Learning multiple layers of features from tiny images (2009)
14. LeCun, Y., et al.: Backpropagation applied to handwritten zip code recognition. Neural Comput. **1**(4), 541–551 (1989)

15. Li, D., Wang, J.: FedMD: heterogenous federated learning via model distillation. arXiv preprint arXiv:1910.03581 (2019)
16. Li, T., Sahu, A.K., Talwalkar, A., Smith, V.: Federated learning: challenges, methods, and future directions. arXiv preprint arXiv:1908.07873 (2019)
17. Li, T., Sahu, A.K., Zaheer, M., Sanjabi, M., Talwalkar, A., Smith, V.: Federated optimization in heterogeneous networks. arXiv preprint arXiv:1812.06127 (2018)
18. Li, W., et al.: Privacy-preserving federated brain tumour segmentation. In: Suk, H.-I., Liu, M., Yan, P., Lian, C. (eds.) MLMI 2019. LNCS, vol. 11861, pp. 133–141. Springer, Cham (2019). https://doi.org/10.1007/978-3-030-32692-0_16
19. Li, X., Gu, Y., Dvornek, N., Staib, L., Ventola, P., Duncan, J.S.: Multi-site fMRI analysis using privacy-preserving federated learning and domain adaptation: abide results. arXiv preprint arXiv:2001.05647 (2020)
20. Liang, P.P., Liu, T., Ziyin, L., Salakhutdinov, R., Morency, L.P.: Think locally, act globally: federated learning with local and global representations. arXiv preprint arXiv:2001.01523 (2020)
21. McMahan, H.B., Moore, E., Ramage, D., Hampson, S., et al.: Communication-efficient learning of deep networks from decentralized data. arXiv preprint arXiv:1602.05629 (2016)
22. Pillutla, K., Kakade, S.M., Harchaoui, Z.: Robust aggregation for federated learning. arXiv preprint (2019)
23. Rieke, N., et al.: The future of digital health with federated learning. arXiv preprint arXiv:2003.08119 (2020)
24. Sattler, F., Müller, K.R., Samek, W.: Clustered federated learning: model-agnostic distributed multi-task optimization under privacy constraints. arXiv preprint arXiv:1910.01991 (2019)
25. Sattler, F., Wiedemann, S., Müller, K.R., Samek, W.: Robust and communication-efficient federated learning from Non-IID data. IEEE Trans. Neural Netw. Learn. Syst. (2019)
26. Sheller, M.J., et al.: Federated learning in medicine: facilitating multi-institutional collaborations without sharing patient data. Sci. Rep. 10(1), 1–12 (2020)
27. Sheller, M.J., Reina, G.A., Edwards, B., Martin, J., Bakas, S.: Multi-institutional deep learning modeling without sharing patient data: a feasibility study on brain tumor segmentation. In: Crimi, A., Bakas, S., Kuijf, H., Keyvan, F., Reyes, M., van Walsum, T. (eds.) BrainLes 2018. LNCS, vol. 11383, pp. 92–104. Springer, Cham (2019). https://doi.org/10.1007/978-3-030-11723-8_9
28. Simonyan, K., Zisserman, A.: Very deep convolutional networks for large-scale image recognition. arXiv preprint arXiv:1409.1556 (2014)
29. Smith, V., Chiang, C.K., Sanjabi, M., Talwalkar, A.S.: Federated multi-task learning. In: Advances in Neural Information Processing Systems, pp. 4424–4434 (2017)
30. Tschandl, P., Rosendahl, C., Kittler, H.: The ham10000 dataset, a large collection of multi-source dermatoscopic images of common pigmented skin lesions. Sci. Data 5, 180161 (2018)
31. Xiao, H., Rasul, K., Vollgraf, R.: Fashion-MNIST: a novel image dataset for benchmarking machine learning algorithms. arXiv preprint arXiv:1708.07747 (2017)
32. Xu, J., Wang, F.: Federated learning for healthcare informatics. arXiv preprint arXiv:1911.06270 (2019)
33. Zhao, Y., Li, M., Lai, L., Suda, N., Civin, D., Chandra, V.: Federated learning with Non-IID data. arXiv preprint arXiv:1806.00582 (2018)

Weight Erosion: An Update Aggregation Scheme for Personalized Collaborative Machine Learning

Felix Grimberg$^{(\boxtimes)}$ iD, Mary-Anne Hartley iD, Martin Jaggi iD,
and Sai Praneeth Karimireddy iD

Ecole polytechnique fédérale de Lausanne, Lausanne, Switzerland
`felix.grimberg@hotmail.com`

Abstract. Background. In medicine and other applications, the copying and sharing of data is impractical for a range of well-considered reasons. With federated learning (FL) techniques, machine learning models can be trained on data spread across several locations without such copying and sharing. While good privacy guarantees can often be made, FL does not automatically incentivize participation and the resulting model can suffer if data is non-identically distributed (non-IID) across locations. Model personalization is a way of addressing these concerns. **Methods.** In this study, we introduce Weight Erosion: an SGD-based gradient aggregation scheme for personalized collaborative ML. We evaluate this scheme on a binary classification task in the Titanic data set. **Findings.** We demonstrate that the novel Weight Erosion scheme can outperform two baseline FL aggregation schemes on a classification task, and is more resistant to over-fitting and non-IID data sets.

1 Background

In medicine and other applications, data is siloed for a range of well-considered reasons including confidentiality and governmental regulations. This creates inequitable access to probabilistic medicine, where those with less accessible or smaller silos are under-represented in medical literature. Additionally, "ownership" of potential scientific results creates an environment of competition in which researchers are reluctant to share their intellectual property (IP). Moreover, delays in granting access can render time-sensitive epidemiological data substantially less relevant and can lead to outdated and poorly adaptive models.

Federated learning (FL) can transform access to highly sensitive data by jointly training a machine learning (ML) model through collaboration across silos without copying the data onto a central server [11]. However, it does not automatically protect IP, and blindly learning a single global model on possibly non-identically distributed (non-IID) data risks creating uninformative insights. One way of addressing these concerns is model personalization, e.g. through *Featurization, Multi-task learning*, or *Local fine-tuning* [7,9,10,14].

© Springer Nature Switzerland AG 2020
S. Albarqouni et al. (Eds.): DART 2020/DCL 2020, LNCS 12444, pp. 160–169, 2020.
https://doi.org/10.1007/978-3-030-60548-3_16

The *Weight Erosion* scheme presented here optimizes a personalized model for one silo, as opposed to conventional personalized FL methods which train personalized models for all silos simultaneously. While it is conceptually related to local fine-tuning, the *Weight Erosion* scheme is optimized to discard contributions from unhelpful[1] silos as early as possible in the training process.

2   Setting and Objective

We consider a network of agents i collecting samples from the underlying distributions $\mathcal{D}_i$. One agent (agent 0) is called the user and wishes to perform an inference task on $\mathcal{D}_0$. The task of *personalized collaborative ML*, in a broad sense, is to give agent 0 a training algorithm which discriminates between the available agents in some way to minimize the true loss of the resulting model on $\mathcal{D}_0$.

Example Clinical Setting. The agents could be individual hospitals dispersed across one or several regions. The aetiology of common, generic symptoms such as fever is highly dependent on geographic location. For instance, rural populations suffer more vector borne diseases such as malaria, while fevers in the urban setting tend to be related to respiratory disease. A clinician (agent 0) might want to train a ML model to help diagnose the patients admitted to their urban hospital. The number N_0 of samples collected at their hospital, however, is too limited to yield a satisfactory model, so agent 0 considers using FL to leverage data from other hospitals. Knowing that rural populations suffer from fairly different problems than agent 0's urban patients, agent 0 pre-selects only other urban hospitals – in a way, agent 0 performs manual model personalization based on prior medical knowledge. However, agent 0 suspects the presence of other confounding variables that could cause the samples collected by *some* of the other urban hospitals to negatively affect the diagnostic accuracy of the trained model on new patients treated by agent 0.

For personalized collaborative ML, formally, we are given:

- A set of agents $i \in \mathcal{U} = \{0, 1, \ldots, N\}$
  - Each agent i has collected a set of samples (called *data set*), on the domain $\mathcal{X} \times \mathcal{Y}$: $\mathcal{S}_i = \left\{ \mathbf{x}_i^{(n)}, y_i^{(n)} \right\}_{n=1,\ldots,N_i}$, each from an (unknown) underlying distribution $\mathcal{D}_i$: $\left(\mathbf{x}_i^{(n)}, y_i^{(n)} \right) \overset{i.i.d.}{\sim} \mathcal{D}_i$.
  - We assume that the label y_0 is not independent of the features $\mathbf{x}_0$ under $\mathcal{D}_0$: $p_0\left(y|\mathbf{x}\right) \neq p_0\left(y\right)$
- A class of models $\mathcal{M}$ s.t. $f : \mathcal{X} \to \mathcal{Y} \; \forall f \in \mathcal{M}$. For instance, $\mathcal{M}$ could be the class of linear models where $f\left(\mathbf{x}\right) = \mathbf{w}^\top \mathbf{x}$, $\mathbf{w} \in \mathbb{R}^D$.

---

[1] The samples stored in other silos can be unhelpful for various reasons. One such reason is intentionally byzantine behaviour, which can be successfully addressed with existing byzantine-robust FL methods [5] and local fine-tuning. However, this approach does not cover cases where silos contain data sampled from a different distribution as laid out in Sect. 2.

– A loss function: $\ell(y, \hat{y})$. For instance, the loss function could be the mean squared error (MSE).

We measure the performance of a model $f \in \mathcal{M}$ by its *expected loss* on $\mathcal{D}_0$, defined in Eq. 1.

$$\mathcal{L}_{\mathcal{D}_0}(f) = \mathbb{E}_{(\mathbf{x},y) \sim \mathcal{D}_0}\left[\ell(y, f(\mathbf{x}))\right] \tag{1}$$

Our goal is to find a training algorithm $\mathcal{A}$ which, given $\mathcal{U}$, $\mathcal{M}$, and ℓ, will produce the model $f = \mathcal{A}(\mathcal{U}, \mathcal{M}, \ell) \in \mathcal{M}$ which minimizes the personalized true loss $\mathcal{L}_{\mathcal{D}_0}(f)$ on $\mathcal{D}_0$ (Eq. 2). This is opposed to standard federated learning methods $\mathcal{A}_{standard}^{FL}$, which consider that all non-malicious agents have samples from the same underlying distribution. Applying such a method in the present setting would minimize the loss on a weighted sum of the agents' distributions $\mathcal{D}_i$, $i \in \mathcal{U}$ instead, as shown in Eq. 3. The weights λ_i depend on the specific learning method, but they are often either $\lambda_i = \frac{1}{|\mathcal{U}|} \forall i \in \mathcal{U}$ or $\lambda_i = \frac{|\mathcal{S}_i|}{\sum_{i \in \mathcal{U}} |\mathcal{S}_i|}$.

Personalized coll. ML: $find:$ $\mathcal{A} = \underset{\mathcal{A}'}{\arg\min} \mathcal{L}_{\mathcal{D}_0}(\mathcal{A}'(\mathcal{U}, \mathcal{M}, \ell)) \tag{2}$

Standard FL result: $\mathcal{A}_{standard}^{FL} = \underset{\mathcal{A}'}{\arg\min} \sum_{i \in \mathcal{U}} \lambda_i \mathcal{L}_{\mathcal{D}_i}(\mathcal{A}'(\mathcal{U}, \mathcal{M}, \ell)) \tag{3}$

3   The Weight Erosion Aggregation Scheme

We present a novel adaptation of federated training algorithms based on *robust aggregation rules* such as in [1,5,13]. Briefly, each agent i is initially given a weight $\alpha_i^0 = 1$. At each round, they compute a mini-batch gradient[2] $\mathbf{g}_i$ and the relative distance $d_{i,0}^{rel}$ between $\mathbf{g}_i$ and $\mathbf{g}_0$ (Eq. 4). The weight α_i is then decreased by a small amount that depends on the distance $d_{i,0}^{rel}$ (Eq. 5). The weights α_i are finally used to compute a personalized weighted average of the gradient vectors $\mathbf{g}_i$ (Eq. 6).

$$d_{i,0}^{rel} = \frac{\|\mathbf{g}_i - \mathbf{g}_0\|}{\|\mathbf{g}_0\|} \geq 0 \tag{4}$$

$$\alpha_i^r = \max\left\{0, \alpha_i^{r-1} - \left(1 + p_s \left\lfloor \frac{(r-1)b}{|\mathcal{S}_i|} \right\rfloor\right) p_d \cdot d_{i,0}^{rel}\right\} \tag{5}$$

$$\bar{\mathbf{g}} \leftarrow \frac{\sum_{i \in \mathcal{U}} \alpha_i^r \mathbf{g}_i}{\sum_{i \in \mathcal{U}} \alpha_i^r} \tag{6}$$

As the number of rounds increases, samples from smaller data sets will be seen more often than samples from larger data sets, because each agent uses the same number of samples per round. To balance this over-representation, the change in α_i (Eq. 5) is made to depend also on the average number of times each sample in $\mathcal{S}_i$ has been used. In Eq. 5, b stands for the batch-size, while p_d is the distance penalty factor and p_s is the size penalty factor. The values of

---

[2] Alternatively, the agents can compute local SGD updates, used in the same way.

Algorithm 1: WEIGHT EROSION

Data: A set of agents $i \in \mathcal{U}$, with associated data sets $\mathcal{S}_i$, a model class $\mathcal{M}$, and a gradient-based machine learning algorithm $\mathcal{A}$.

Result: A personalized machine learning model f.

Set a number of rounds r_{max}, a batch size b, a distance penalty factor p_d, and a size penalty factor p_s ;

Initialize $\alpha_i^0 \leftarrow 1 \; \forall i \in \mathcal{U}$;

Randomly initialize the model $f \in \mathcal{M}$;

for r *from* 1 *to* r_{max} **do**

    for $i \in \mathcal{U}$, *starting with* $i = 0$ **do**

        Select a batch of size b from $\mathcal{S}_i$ and compute a gradient $\mathbf{g}_i$;

        Compute the distance $d_{i,0}^{rel}$ (Equation 4) and α_i^r (Equation 5) ;

    end

    $\bar{\mathbf{g}} \leftarrow \frac{\sum_{i \in \mathcal{U}} \alpha_i^r \mathbf{g}_i}{\sum_{i \in \mathcal{U}} \alpha_i^r}$;

    Update the model f based on $\bar{\mathbf{g}}$;

end

---

these penalty factors should be picked by the user, as they control the degree of personalization. Selecting a large value for p_d leads to a rapid decay of all agent weights, limiting the collaborative scope of the training. Conversely, if p_d is very low, available agents' contributions are used equally regardless of their $d_{i,0}^{rel}$. Finally, p_s controls how strongly small data sets are penalized to counteract their over-representation.

We next discuss some salient properties of our algorithm.

Privacy and Robustness. Weight Erosion can be seamlessly integrated into the privacy-preserving FL protocol proposed in [5], which leverages secure multiparty computations (MPC) to obtain distances while keeping the individual gradients private at all times. Replacing the Byzantine robust aggregation rule by Weight Erosion maintains the strong privacy guarantees, together with a weaker form of robustness since any Byzantine agent i would likely see their weight α_i decline very fast.

Incentives and IP. Agents could be rewarded for their data collection efforts according to how many rounds they participate in, or according to their weight (summed over all rounds). Intuition tells us that the result of Algorithm 1 lies somewhere between the global model and agent 0's local model (cf. Appendix A). Each agent i can therefore verify before participating, that the other agents contribute enough data to dilute the influence of $\mathcal{S}_i$ and protect agent i's IP.

Interpretability. With this approach, the user does not learn why a certain subset of the available data sets were selected, even if they understand the selection process. Indeed, the selection is only revealed to them *after* the model has been trained.

Resilience. Like [5], Algorithm 1 does not break if users appear or disappear between subsequent rounds. When a new agent appears after the first round, their weight should not be initialized to 1, but rather to the mean or median weight of all agents.

4   Application Case Study

We set up a collaborative ML simulation using the publicly available Titanic data set [3]. This set collects 14 features on passengers of the cruise ship Titanic, including their survival status. We train a prediction model for the survival status of passengers based on the following features: *fare*, *passenger class*, *port of embarkation*, travelling *alone or accompanied*, *sex*, *age*, and being either *adult or minor* (aged 16 or less). The pre-processing procedure is largely aligned with [12].

Implementation. A generic `Python` framework for efficient FL simulations on a single machine with `JAX` and `Haiku` [2,6] was created and made publicly available by the authors [8]. The Weight Erosion aggregation rule (Algorithm 1) is implemented in an altered version of this framework. All code (including the pre-processing) is made publicly available on GitHub [4]. Due to the small data set size, the number of agents is limited by splitting it into four subsets (agents 0–3) based on the variable *age*. The four-way split is undertaken in two ways to simulate IID and non-IID settings, as follows:

`AGE_STRICT`: Samples are strictly segregated into groups based on their age: agent 0: patients aged 0–20 years old / agent 1: 21–35 years old / agent 2: 36+ years old / agent 3: age unknown. In this data set, the *age*, *adult or minor* and some other features are strongly correlated with age. This feature skew should be sufficient to affect the conditional probability $p(y|\mathbf{x})$.

`AGE_SOME`: Agents 0 and 1 randomly partition the patients aged 0–35 years old among themselves, while agent 2 has all patients aged 36+ years old / agent 3: age unknown. This should make agents 0 and 1 more helpful for each other than agents 2 and 3.

5   Results

The collaborative training of a prediction model by four agents is simulated on a single machine for the data sets and splits detailed in Sect. 4. Each simulation is repeated four times, such that each agent serves as a user once. Every time, the user's data set is split into a test set and a training set of equal sizes, whereas 100% of the other agents' data sets are used for training. A classification model, consisting of a 2-output linear regression layer followed by log-softmax, is trained using federated SGD with three different gradient aggregation schemes:

`FedAvg`:   Train on all agents' training sets without Weight Erosion.
`Weight Erosion`:   Train on all agents' training sets with Weight Erosion.
`Local`:   Train only on the user's training set.

At each communication round, each model's accuracy is measured on the user's test set and reported (along with the weights α_i in the Weight Erosion scheme).

5.1   Predicting Survival of the Titanic with AGE_STRICT Split

In Fig. 1 (top), we can see the compensating effect of p_s at work, as agent 1's weight α_1 experiences the slowest decrease in all three graphs because agent 1 has collected one batch more than the other agents. Nevertheless, the difference between α_1 and the other agent weights is only very marked when agent 2 is the user, showing that the impact of p_d rivals that of p_s.

When splitting the Titanic data set into agents by age, we notice that the survival rate (Fig. 1 top: lower of the two red dotted lines in each graph) is fairly similar across all age groups. This implies that the *age* feature may not be particularly useful in predicting the survival of passengers. However, the survival rate is much lower among the group of passengers whose age is unknown. This could simply be a sampling bias, as it would seem easier to research the age of survivors than that of the deceased.

Another debatable feature is whether the passenger is a child (Passenger is minor), which is false across all data of agents 1 and 2, and unknown in the data set of agent 3. We observe that agent 0 stands out by the fact that the Weight Erosion model clearly out-performs the Local model. Nevertheless, the difference in prediction accuracy does not come from the Local model overfitting on the Passenger is minor feature, since its weights are nearly identical in both models as we see in Fig. 2 (top left, top center).

Travelling alone is strongly correlated with death in both models, indicating that passengers aged 0–20 years old were more likely to die when travelling alone. Notably, the same feature is weakly correlated with survival in the FedAvg model, as well as in the very accurate Local and Weight Erosion models trained with agent 2 as user (cf. Fig. 2: bottom left, bottom center).

In Fig. 2 (top left, top center), three features have smaller weights in the Weight Erosion model than in the Local model: Travelling alone, boarding in Queenstown (as opposed to Southampton or Cherbourg), and travelling in Second Class. These differences could be a sign that the local model is overfitting, especially if they apply to a small portion of the passengers, as is the case with Queenstown and presumably Travelling alone.

Interestingly, the Weight Erosion aggregation scheme can produce a model that is not merely a weighted average of the FedAvg and Local models. Indeed, in Fig. 2 (top), the weights of some features such as Sex, Age, or boarding in Queenstown, differ more between the FedAvg model and the Weight Erosion model, than they do between the FedAvg model and the Local model.

In Fig. 2 (bottom left, bottom center), we investigate the weights of the models trained with Local and Weight Erosion schemes when agent 2 is the user, since these models perform exceptionally well (> 90%) despite the unexceptional survival probability of 40%. We observe that weights with absolute values < 0.25 differ between the models, without affecting their performance, while larger weights are very consistent. Further, two main features stand out: Firstly, the weights for Sex are spectacularly large (with absolute values between 1.1 and 1.4). Given the models' extraordinary test accuracies, we conclude that most survivors aged 36+ were female. Secondly, the weights of the First Class and

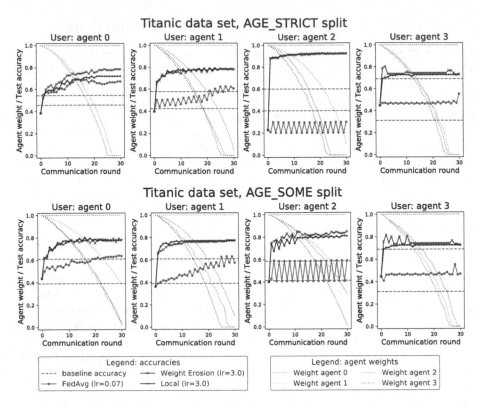

Fig. 1. Predicting survival of the Titanic. The data set is split across 4 agents by age. Full lines represent each model's accuracy on the user's test set. Displayed in red (dashed) is the accuracy obtained by predicting always 1 or always 0. The learning rates were tuned independently for each aggregation scheme (in legend: lr). Training on one batch per round and agent. $p_d = 0.01$, $p_s = 0.2$, seed = 278.
Top: AGE_STRICT split. Batch size: 161 (agent 1: 3 batches, others: 2 each).
Bottom: AGE_SOME split. Batch size: 132 (agent 3: 2 batches, others: 3 each).

Second Class features show that, in this age group, only First-Class passengers were much more likely to survive than Third-Class passengers.

5.2 Predicting Survival of the Titanic with AGE_SOME Split

As expected, we observe that agents 0 and 1, whose samples are drawn from the same age group, have the highest weight in each other's model, while their weights are similar to each other in the two other agents' models (Fig. 1, bottom). The models trained by these agents achieve the same test accuracy as with the AGE_STRICT split, save for the spectacular accuracy obtained for agent 2 in Fig. 1 (top).

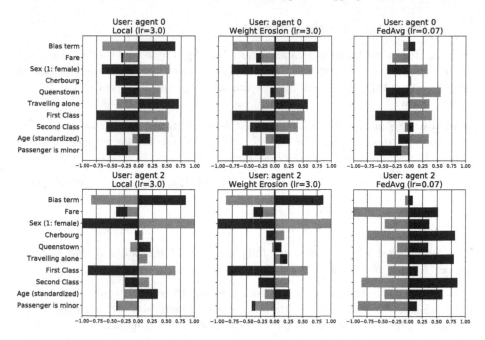

Fig. 2. Top: Parameters of the models trained in Fig. 1 (top left). Bottom: Parameters of the models trained in Fig. 1 (top, agent 2). Black: Weights for the first output of the linear layer. Gray: Weights for the second output, e.g.:

: The feature is strongly correlated with death.

: The feature is strongly correlated with survival.

6   Conclusion

In this paper, we introduce a novel method of model personalization in collaborative ML: Weight Erosion. Our application case study demonstrates that it can outperform two baseline schemes (FedAvg and local training), by converging to a better model that is not a linear combination of the local and global models.

Further Work. Additional refinement is needed to address in a more equitable way the under-representation of samples from larger data sets, e.g. by introducing a different weight β_i that is unrelated to α_i. Equally importantly, the Weight Erosion scheme should be tested on different ML tasks, such as image classification. For instance, the need for model personalization could arise as clinicians train a risk stratification model for COVID-19 pneumonia to identify pathological patterns in lung ultrasound images that warrant hospitalization. Personalized collaborative ML would allow them to leverage existing lung ultrasound acquired in other hospitals without being negatively affected by geographic variations in aggravating factors (such as diabetes or exposure to tuberculosis and indoor air pollution), nor by other confounding variables. Additionally, the patterns of pathology that warrant hospitalization may be assessed differently in light of local resources and thus vary between subsets of agents.

A   Analyzing Weight Erosion

Let us analyze what happens in a few examples, if $\mathbf{g}_i$ are indeed the full gradients (i.e., computed on the entire set $\mathcal{S}_i$), as opposed to stochastic gradients. Let us further assume that each training loss function $\mathcal{L}_{\mathcal{S}_i}(f)$ is convex in the model parameters $\mathbf{w}$.

Example 1. *Suppose the model has just been randomly initialized to f^0 and performs very sub-optimally for both $\mathcal{S}_i$ and $\mathcal{S}_0$. Then, the normalized distance between the gradients will be very low. Formally:*

$$\text{Let}\quad \min\left\{\mathcal{L}_{\mathcal{S}_0}\left(f^0\right),\mathcal{L}_{\mathcal{S}_i}\left(f^0\right)\right\} \gg \max\left\{\mathcal{L}_{\mathcal{S}_0}\left(f_{\mathcal{S}_i}\right),\mathcal{L}_{\mathcal{S}_i}\left(f_{\mathcal{S}_0}\right)\right\},\quad f_{\mathcal{S}_i}=\underset{f\in\mathcal{M}}{\arg\min}\ \mathcal{L}_{\mathcal{S}_i}\left(f\right)$$

$$\Rightarrow\qquad\qquad \|\mathbf{g}_i-\mathbf{g}_0\| \ll \|\mathbf{g}_0\|$$

$$\Rightarrow\qquad\qquad d_{i,0}^{rel} \ll 1$$

Therefore, both agents i and 0 are fully included at this stage of the training process, because the distance $d_{i,0}^{rel}$ is close to 0.

However, as the training process progresses and the model gradually performs better, the distance between the gradients steadily increases. We can analyze the edge cases where f is either the global model or the local model:

Example 2. *Suppose f is the global model for 2 agents, 0 and i:*

$$\text{Let}\qquad f=\underset{f'\in\mathcal{M}}{\arg\min}\ \mathcal{L}_{\mathcal{S}_0\cup\mathcal{S}_i}\left(f'\right),\qquad \mathcal{L}_{\mathcal{S}_0\cup\mathcal{S}_i}\left(f\right)=\frac{|\mathcal{S}_0|\mathcal{L}_{\mathcal{S}_0}\left(f\right)+|\mathcal{S}_i|\mathcal{L}_{\mathcal{S}_i}\left(f\right)}{|\mathcal{S}_0|+|\mathcal{S}_i|}$$

$$\Rightarrow\qquad \mathbf{0}=\nabla_{\mathbf{w}}\mathcal{L}_{\mathcal{S}_0\cup\mathcal{S}_i}\left(f\right)=\frac{|\mathcal{S}_i|}{|\mathcal{S}_0|+|\mathcal{S}_i|}\left(\frac{|\mathcal{S}_0|}{|\mathcal{S}_i|}\mathbf{g}_0+\mathbf{g}_i\right)$$

$$\Rightarrow\qquad \mathbf{g}_i=-\frac{|\mathcal{S}_0|}{|\mathcal{S}_i|}\mathbf{g}_0$$

$$\Rightarrow\qquad d_{i,0}^{rel}=\frac{\|\mathbf{g}_i-\mathbf{g}_0\|}{\|\mathbf{g}_0\|}=\frac{\left\|\left(-\frac{|\mathcal{S}_0|}{|\mathcal{S}_i|}-1\right)\mathbf{g}_0\right\|}{\|\mathbf{g}_0\|}=1+\frac{|\mathcal{S}_0|}{|\mathcal{S}_i|}$$

In this case, whether (and to which degree) agent i should participate in the training depends on the sizes $|\mathcal{S}_0|$ and $|\mathcal{S}_i|$: Indeed, if $|\mathcal{S}_0| \ll |\mathcal{S}_i|$, then $d_{i,0}^{rel} \approx 1$, indicating that it would be useful to incorporate agent i further in the training process. This is sensible, considering that the much larger number of samples in $\mathcal{S}_i$ could help reduce the generalization error substantially. Inversely, if $|\mathcal{S}_0| \gg |\mathcal{S}_i|$, then it is not useful to include $\mathcal{S}_i$ in training and we are better off only using the (much more numerous) samples collected by agent 0. Correspondingly, this leads to $d_{i,0}^{rel} \gg 1$.

Example 3. *Suppose now that f is the local model of agent 0:*

$$\text{Let}\qquad f=\underset{f'\in\mathcal{M}}{\arg\min}\ \mathcal{L}_{\mathcal{S}_0}\left(f'\right)$$

$$\Rightarrow\qquad \mathbf{g}_0=\nabla_{\mathbf{w}}\mathcal{L}_{\mathcal{S}_0}\left(f\right)=\mathbf{0}$$

$$\Rightarrow\quad \forall i\in\mathcal{U}:d_{i,0}^{rel}\ \text{undefined}\ (+\infty)$$

Under the stated assumptions of convexity and full gradient descent, all distances $d_{i,0}^{rel}$ grow without an upper bound as the trained model approaches the local model. In other words, when we move too close to the local model, all data sets start to appear very different from S_0. Consequently, using any decreasing function of $d_{i,0}^{rel}$ as a similarity metric results in the danger of converging to the local model. One possible strategy to prevent this is to stop training before convergence. Intuitively, training should not be stopped as long as the gradients fit into a D-dimensional cone[3].

References

1. Blanchard, P., Guerraoui, R., Stainer, J., et al.: Machine learning with adversaries: byzantine tolerant gradient descent. In: Advances in Neural Information Processing Systems, pp. 119–129 (2017)
2. Bradbury, J., et al.: JAX: composable transformations of Python+NumPy programs (2018). http://github.com/google/jax
3. Harrell Jr., F.E.: Titanic dataset, October 2017. https://www.openml.org/d/40945
4. Grimberg, F., Jaggi, M.: Semester project on private and personalized ml (2020). https://github.com/epfl-iglobalhealth/coML-Personalized-v2
5. He, L., Karimireddy, S.P., Jaggi, M.: Secure byzantine-robust machine learning. arXiv:2006.04747 [cs, stat] (2020). http://arxiv.org/abs/2006.04747
6. Hennigan, T., Cai, T., Norman, T., Babuschkin, I.: Haiku: Sonnet for JAX (2020). http://github.com/deepmind/dm-haiku
7. Kairouz, P., McMahan, H.B., Avent, B., et al.: Advances and open problems in federated learning. arXiv:1912.04977 [cs, stat] (2019). http://arxiv.org/abs/1912.04977
8. Karimireddy, S.P.: Jax federated learning (2020). https://tinyurl.com/Karimireddy-Jax-FL
9. Mansour, Y., Mohri, M., Ro, J., Suresh, A.T.: Three approaches for personalization with applications to federated learning. arXiv:2002.10619 [cs, stat] (2020). http://arxiv.org/abs/2002.10619
10. Mansour, Y., Mohri, M., Rostamizadeh, A.: Domain adaptation: learning bounds and algorithms. arXiv preprint arXiv:0902.3430 (2009)
11. McMahan, H.B., Moore, E., Ramage, D., Hampson, S., Agüera y Arcas, B.: Communication-efficient learning of deep networks from decentralized data. In: Proceedings of the 20th International Conference on Artificial Intelligence and Statistics (AISTATS). http://arxiv.org/abs/1602.05629 (2016)
12. MNassri, B.: Titanic: logistic regression with python (2020). https://www.kaggle.com/mnassrib/titanic-logistic-regression-with-python?scriptVersionId=26445092
13. Pillutla, K., Kakade, S.M., Harchaoui, Z.: Robust aggregation for federated learning. arXiv preprint arXiv:1912.13445 (2019)
14. Smith, V., Chiang, C.K., Sanjabi, M., Talwalkar, A.S.: Federated multi-task learning. In: Advances in Neural Information Processing Systems, pp. 4424–4434 (2017)

---

[3] If all gradients fit into a cone, then it is possible to improve the loss on all data sets by taking a step in the opposite direction of the axis of the cone.

Federated Gradient Averaging for Multi-Site Training with Momentum-Based Optimizers

Samuel W. Remedios[1,2,3(✉)], John A. Butman[3], Bennett A. Landman[2], and Dzung L. Pham[3,4]

[1] Johns Hopkins University, Baltimore, MD 21218, USA
[2] Vanderbilt University, Nashville, TN 37235, USA
[3] Clinical Center, National Institutes of Health, Bethesda, MD 20814, USA
samuel.remedios@nih.gov
[4] Center for Neuroscience and Regenerative Medicine, Henry M. Jackson Foundation, Bethesda, MD 20817, USA

Abstract. Multi-site training methods for artificial neural networks are of particular interest to the medical machine learning community primarily due to the difficulty of data sharing between institutions. However, contemporary multi-site techniques such as weight averaging and cyclic weight transfer make theoretical sacrifices to simplify implementation. In this paper, we implement federated gradient averaging (FGA), a variant of federated learning without data transfer that is mathematically equivalent to single site training with centralized data. We evaluate two scenarios: a simulated multi-site dataset for handwritten digit classification with MNIST and a real multi-site dataset with head CT hemorrhage segmentation. We compare federated gradient averaging to single site training, federated weight averaging (FWA), and cyclic weight transfer. In the MNIST task, we show that training with FGA results in a weight set equivalent to centralized single site training. In the hemorrhage segmentation task, we show that FGA achieves on average superior results to both FWA and cyclic weight transfer due to its ability to leverage momentum-based optimization.

Keywords: Federated learning · Multi-site · Deep learning

1 Introduction

The quality of a machine learning model stems directly from its training data. Deep neural networks especially benefit from large datasets, utilizing thousands or millions of parameters to learn salient feature weights to account for many types of variation. The same is true when applying such models to medical imaging, except that site effects such as scanner manufacturer, acquisition parameters, and subject cohort greatly impact the generalization of a model trained to different sites. It is common to find models trained from a single site's data

© Springer Nature Switzerland AG 2020
S. Albarqouni et al. (Eds.): DART 2020/DCL 2020, LNCS 12444, pp. 170–180, 2020.
https://doi.org/10.1007/978-3-030-60548-3_17

perform poorly on a new site [7]. This problem is further compounded by the difficulty of data sharing with medical images, which arises from ethical, logistical, and political concerns.

To address this problem, several multi-site approaches have been suggested. The term "federated learning" (FL) was first widely known from [12] in the application of learning from mobile devices. Here, the authors introduced a scenario in which an arbitrary number of clients solve individual optimization problems and aggregate their parameters as a central weighted average. This approach to FL has already seen implementation on real-world medical datasets such as brain tumor segmentation [14]. Similarly, asynchronous stochastic gradient descent [36] considered different GPU devices as clients, splitting training data among them and aggregating gradients before performing weight update to decrease training time. Related research in the field of continual learning [11, 20, 20] investigates methods to mitigate catastrophic forgetting [6], the phenomenon by which neural networks have no guarantee to retain useful weight values when training on new data. Bayesian neural networks, in which parameters are not point estimates but distributions of parameters, allow techniques such as distributed weight consolidation [18] to consider multi-site training as a continual learning problem. In transfer learning [22, 34], also known as pre-training or fine-tuning, model parameters are learnt first on some (usually large) dataset and subsequently trained for the problem of interest on a smaller dataset. Cyclic weight transfer (CWT) [4, 25] is a distributed method that leverages transfer learning to iteratively train a model at different institutes. While model generalization was reported to have improved, there is no mathematical guarantee that the CWT models will converge to an optimum of the combined datasets.

Although recent works [15] have shown the theoretical convergence for federated weight averaging (FWA, previously formulated as FedAVG [19]), in practice there is still a performance gap between centralized training at a single site and collaborative learning methods [28]. In contrast, federated gradient averaging (FGA) should offer equivalent performance to single-site learning, first formulated as FedSGD [19]. Gradient averaging has been a standard practice since the advent of deep learning [3] with regards to training a neural network using batches, as weight updates from a single sample are noisy and updates from the entire dataset are undesirable for empirical and theoretical reasons. Gradient averaging has also been used in asynchronous stochastic gradient descent [36] and in the TernGrad work [32], which considers the quantization of gradients to reduce network overhead.

In this work, we recast FedSGD [19] as FGA, show its equivalence to centralized training, and show FGA permits the use of not only SGD but also momentum-based optimizers such as Adam [10] without sacrificing the benefits of momentum via variable resets. We show the first use case of FGA for learning from unshared data housed at physically separate institutes and directly compare it to federated weight averaging (FWA) [14] and cyclic weight transfer (CWT) [25]. We first evaluate FGA in a simulated disjoint multi-site scenario with the handwritten digit classification dataset MNIST [13], then on a multi-site CT hemorrhage segmentation task.

2   Method

2.1   Background

The training of artificial neural networks can be interpreted as finding parameters Θ for a function f which minimize a loss function $\mathscr{L}$ over N input-output pairs $\{x_i, y_i\}, i = 1, 2, ..., N$:

$$\arg\min_{\Theta} \sum_i^N \mathscr{L}(f_\Theta(x_i), y_i) \tag{1}$$

These parameters are usually updated iteratively via gradient descent optimization:

$$\Theta_t \leftarrow \Theta_{t-1} + \eta \nabla_{\Theta_{t-1}} \frac{1}{N} \sum_i^N \mathscr{L}(f_{\Theta_{t-1}}(x_i), y_i) \tag{2}$$

In other words, the new weights Θ at time t are updated from the previous weights at time $t - 1$ plus the gradient of the loss function scaled down by η, the learning rate (or step size). It is conventional to use Stochastic Gradient Descent (SGD)-based optimizers in practice, where the gradient is not calculated over all of the training samples but a randomly sampled subset (known as a batch).

To accelerate training, several momentum-based optimizers have been proposed [27]. The use of momentum has been shown to reduce sensitivity to noisy gradients and improve overall convergence. Adam [10] is a ubiquitous deep learning optimizer that updates parameters via momentum variables $\hat{m}$ and $\hat{v}$ (described in Algorithm 1 in [10]), each of which are functions of the gradients $\nabla_{\Theta_{t-1}}$.

$$\Theta_t \leftarrow \Theta_{t-1} + \eta \frac{\hat{m}_t}{\sqrt{\hat{v}_t} + \epsilon} \tag{3}$$

2.2   Federated Learning

Multi-site learning approaches aim to find parameters Θ which minimize the loss L not over just N input-output pairs at a single site, but at all participating sites $s = 1, 2, ..., S$ of S participating sites. The class of multi-site learning approaches that we will discuss make use of a central server c which administrates weight and gradient collection, aggregation, and distribution.

Because the number of samples calculated within a batch is independent of the gradient calculation, the gradient of the average is the average of the gradients. We initialize the model f_Θ at the central server c and copy the model to each local site s. The local sites sample a training batch and compute a gradient, then send the gradient back to the central server. The server averages the gradients across sites and sends the averaged gradient to each local site. Each local site computes its own momentum terms based on the averaged gradient and updates weights. In this way, the momentum across sites is guaranteed to be identical (within floating point errors) due to calculation based on the same gradient. The exact procedure is described in Algorithm 1, expanded from FedSGD [19] to work with momentum-based optimizers.

Algorithm 1. FGA implementation. As in [10], g_t^2 indicates the Hadamard (element-wise) product $g_t \odot g_t$, and β_1^t and β_2^t refer to raising β_1 and β_2 to the power t.

Require: local training data $\mathcal{D}_s = \{x_i, y_i\}_{i=1}^{N_s}$
Require: central learning rate η, local decay rates $\beta_{s,1}, \beta_{s,2}$, global small constant ϵ
Require: differentiable loss function $\mathcal{L}$ defined on training pairs (x, y)
Require: Central initialized model Θ_c
 1: **procedure** GLOBAL_TRAINING
 2:     **for** $s \leftarrow 1, 2, ..., \mathcal{S}$ **do**
 3:         Initialize local site time-step: $t_s \leftarrow 0$
 4:         Initialize local site momentum terms: $m_{s,t} \leftarrow 0, v_{s,t} \leftarrow 0$
 5:         Initialize local site model: $\Theta_{s,t} \leftarrow \Theta_c$
 6:     **end for**
 7:     **while** Convergence criteria not met **do**
 8:         **for** $s \leftarrow 1, 2, ..., \mathcal{S}$ **do**
 9:             $t_s \leftarrow t_s + 1$
10:             Sample local site training batch: $\{x_{s,i}, y_{s,i}\} \in \mathcal{B}_s \sim \mathcal{D}_s$
11:             Compute local site batch gradient: $g_s \leftarrow \nabla_\Theta \mathcal{L}(f_{\Theta_{s,t}}(x_{s,i}), y_{s,i})$
12:             Send gradient to central server
13:         **end for**
14:         Central server averages gradients across sites: $g_c \leftarrow \frac{1}{S} \sum_{s=1}^{S} g_s$
15:         Send averaged gradient g_c to each local client.
16:         **for** $s \leftarrow 1, 2, ..., \mathcal{S}$ **do**
17:             Compute 1st moment: $m_{s,t} \leftarrow \beta_{s,1} \cdot m_{s,(t-1)} + (1 - \beta_{s,1}) \cdot g_c$
18:             Compute 2nd moment: $v_{s,t} \leftarrow \beta_{s,2} \cdot v_{s,(t-1)} + (1 - \beta_{s,2}) \cdot g_c^2$
19:             Compute bias-corrected 1st moment: $\hat{m}_{s,t} \leftarrow m_{s,t}/(1 - \beta_{s,1}^t)$
20:             Compute bias-corrected 2nd moment: $\hat{v}_{s,t} \leftarrow v_{s,t}/(1 - \beta_{s,2}^t)$
21:             Update local model: $\Theta_{s,t} \leftarrow \Theta_{s,(t-1)} - \eta \cdot \hat{m}_{s,t}/(\hat{v}_{s,t} + \epsilon)$
22:         **end for**
23:     **end while**
24: **end procedure**

2.3   Implementation

We have implemented FGA as a set of two main Python scripts: server and client. We enabled communication between server and client with secure shell tunneling. The server script is a Python Flask server which runs indefinitely and awaits incoming gradients. After gradients from all participating sites are received they are averaged element-wise and returned to each client. The client script is identical at each site and is a conventional neural network script with the additional step of sending gradients to the server and awaiting the averaged gradient before using an optimizer to update the local weights. Our source code is publicly available here [24]. FL was conducted between two different physical institutions: the NIH Clinical Center (A) and Vanderbilt University (B). At A, we used TensorFlow 2.1, a Tesla V100-SXM3 GPU, CUDA version 10.1, and NVIDIA driver version 418.116.00. At B, we used TensorFlow 2.0 with the TensorFlow Determinism Patch [21], a GeForce RTX 2080 Ti GPU, CUDA version 10.2, and NVIDIA driver version 440.48.02.

3   Experiments

We evaluated our implementation of FGA in a two scenarios: the handwritten dataset MNIST [13] data with simulated multi-site separation and a real multi-site dataset for CT hemorrhage segmentation [25]. With MNIST, we compare FGA to single site centralized data training, single site training at A and B, FWA, and CWT. With CT hemorrhage segmentation, we make the same comparisons except for centralized data training due to privacy restrictions on data transfer. Because there are two training institutes in these experiments, our implementation of CWT results in two different models; one model begins at institute A and after 100 epochs ends at institute A and the other model begins at institute B and ends at institute B. Both CWT models are evaluated at all test sites.

3.1   MNIST

From the MNIST dataset we split the training and validation sets into two disjoint sets such that A has image-label pairs of digits $0, 1, 2, 3, 4$ and B has image-label pairs of digits $5, 6, 7, 8, 9$. Models at both sites have 10 final neurons and thus have the capacity to predict classes at the other site, but have no supporting training or validation data. The test set is entirely withheld and contains samples from all ten digits. An epoch was defined as the model training on every sample from the training set once.

A small convolutional neural network (CNN) architecture was used, consisting of four convolutional layers with 16 3×3 filters activated by ReLU alternating with max pooling layers and finally connected through global max pooling to a fully connected layer with neurons equal to the number of classes. Softmax activation was used for final classification probabilities. All input and output data as well as model weights were 64-bit floats for greater precision in comparing FGA and single-site centralized data training. Convergence was defined as completing 100 training epochs. We trained the CNN using the Adam optimizer with a learning rate of 10^{-4} and with a batch size of 4096 for all methods except FGA for which the client batch size was 2048 to compare against single-site centralized-data training. This is because FGA updates weights by gradients from all participating sites. As a result, only 6 batches were needed to complete an epoch with FGA. To facilitate comparison between FGA and single-site centralized-data training, training data were not shuffled between epochs.

3.2   Hemorrhage Segmentation

A full description of the CT head imaging data is provided in [25]. Site A consisted of 34 training volumes and 34 testing volumes, and site B had 32 training volumes and 27 testing volumes. Additional test data (stored at Site A but acquired at different locations and using different scan protocols) were labeled Site C (11 volumes) and Site D (20 volumes). Since all selected multi-site methods are agnostic to architecture choice, a modified, reduced U-net [26] was chosen as the CNN architecture, replacing double convolution layers and max pooling layers with single convolution layers of kernel size 5 and stride 2 on the down-sampling arc and transpose convolution layers with the same kernel and stride settings to replace double convolution and up-sampling layers on the up-sampling arc. The exact architecture is publicly available [24]. These implementation choices were made to retain the same field-of-view but reduce memory consumption and processing time. This 2D reduced U-net was trained on randomly extracted 2D patches as described in [25] for 100 epochs with a batch size of 256 using the Adam optimizer with a learning rate of 10^{-4}. During training and before patch collection, 20% of the CT volumes are set aside for validation.

4   Results

4.1   MNIST

A comparison of all multi-site learning methods is shown in Table 1. Since our goal is to show the equivalence of FGA to centralized training, the final convergence of the model is secondary to the weight values. When inspecting the individual weight values of single site $A \cup B$ and FGA, they differed by no more than 10^{-12} due to floating point rounding due to gradient calculations. This is expected because FGA is mathematically equivalent to centralized training. Regarding performance across methods, as expected, Single Site A and B each only predict the classes for which training data was provided, resulting in about 50% accuracy. Since CWT methods each terminate at one site, their models are biased towards their most recent site.

4.2   Hemorrhage Segmentation

Qualitative results are shown in Fig. 1 and quantitative results are shown in Table 2. Considering the union of all test sets, FGA significantly ($p < 0.005$) outperforms the other multi-site training methods although the improvement within some test data sets was subtle. On dataset D, cyclic model B achieved greater performance, though FGA is still comparable and more stable, as it does not depend on CWT terminating at a site which may have a similar distribution to that test set. Qualitatively, all methods generated reasonable results, but in general FGA consistently suffered from fewer false positives. CWT performed as well or better than FWA in this scenario, as opposed to the MNIST scenario with disjoint training data.

Table 1. MNIST experiment: balanced accuracy scores (BAS) for each training method and site for the test set. The test set consists of all 10 classes and is identical across sites and methods.

Method	Single Site			Cyclic		FWA	FGA
Train Site	A	B	$A \cup B$	A	B		
BAS	49.39%	48.26%	**93.74%**	79.34%	77.94%	88.37%	**93.74%**

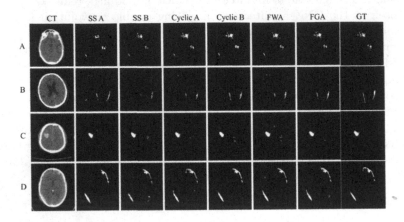

Fig. 1. Qualitative results of segmentation methods at each site.

5   Discussion

We have implemented federated gradient averaging (FGA) with the intent to enable momentum-based optimizers to be used in collaborative learning scenarios without sacrificing the benefits of momentum. We first validated FGA on a disjoint handwritten digit classification task and compared to FWA, CWT, and single-site training. Since FGA results in identical weights at each learning step to the single site scenario, we conclude that they are equivalent. We then applied

Table 2. Dice scores for each training method and site for test sets at each site. Statistical significance with $p < 0.005$ is determined by the paired t-test and is indicated by an asterisk, and best results are indicated by bold text.

Method		Site A	Site B	Site C	Site D	All Sites
Single site	A	0.67 ± 0.06	0.69 ± 0.08	0.69 ± 0.07	0.63 ± 0.05	0.69 ± 0.08
	B	0.71 ± 0.06	0.69 ± 0.08	0.75 ± 0.07	0.64 ± 0.06	0.73 ± 0.07
Cyclic	A	0.72 ± 0.07	0.73 ± 0.07	0.74 ± 0.08	0.63 ± 0.06	0.73 ± 0.07
	B	0.73 ± 0.05	0.72 ± 0.07	0.73 ± 0.08	$\mathbf{0.66 \pm 0.06}$	0.72 ± 0.07
FWA		0.71 ± 0.06	0.72 ± 0.06	0.71 ± 0.08	0.63 ± 0.06	0.72 ± 0.06
FGA		$\mathbf{0.77 \pm 0.07^*}$	$\mathbf{0.76 \pm 0.08^*}$	$\mathbf{0.78 \pm 0.07}$	0.65 ± 0.07	$\mathbf{0.76 \pm 0.08^*}$

FGA to truly multi-site data in a head CT hemorrhage segmentation task and showed its improved performance over other federated methods. Furthermore, we have shown that this approach is stable for application in real-world scenarios where different physically separate sites have different hardware, framework versions, driver versions, and CUDA library versions.

Regarding training time and communication overhead, FGA is about $2\times$ slower than FWA and CWT, and FWA and cyclic weight transfer are about the same speed as the slowest machine in single site training. Regarding scalability, FGA suffers no time penalty as the number of participating sites increases but is hampered in speed by the slowest participating site. CWT does not scale well as the number of participants increases, as the model must complete an epoch at each site before "seeing" all the data. FWA enjoys the best of both worlds, scaling well with the number of participants without delaying learning from unseen sites between epochs, but does not achieve performance equal to centralized training [14,19,28]. However, in general, FGA suffers from much more communication overhead than FWA and CWT, which indeed was partially the motivation for FWA [19]. There are many other factors involved in determining convergence time across sites, including variation in traditional hyperparameter choice (batch, learning rate, number of training epochs), dataset size, hardware, framework and driver versions, and network connectivity. Beyond these, all methods proposed in this paper involve some form of synchronization. For FGA, every step is performed in lockstep, and thus all sites must wait for the slowest site to finish processing its batch before averaging is performed. For FWA, every epoch is performed in lockstep and all sites must therefore wait for the slowest site to complete its epoch before performing the average. Additionally, many methods have considered gradient averaging as incurring too much network overhead, which is one reason why weight averaging is preferred [19,35]. There is also recent research to transform the gradients into communication-efficient variants [2,16,32,33].

Finally, when deploying collaborative learning the protection of patient privacy is of utmost importance, especially when working with medical data [1,5,17, 31]. Previous works have demonstrated the dangers of sending raw data through the network [30] and to this end differentially private (DP) methods have been proposed [8,14,29]. With both FWA and FGA approaches, implementation of DP mitigates one class of privacy concerns, although FWA benefits particularly from obfuscation of batch gradients by sending weights only at the end of the epoch instead of at the end of the batch. In this sense, FGA relies even more on DP methods. However, the reconstruction of training data from the weights or gradients alone is nontrivial [30]. Despite difficulties to reconstruct data or determine whether a subject was a participant in a study, all collaborative multi-site learning approaches are still vulnerable to malicious participant attacks [9]. Even DP models may leak training distribution information since they operate by tracking weight update changes over time. Best practices regarding safeguarding of patient data, model variants, and collaborative learning method selection are still under discussion by groups such as Project MONAI [23].

Acknowledgments. We would like to first thank Shuo Han, Blake Dewey, Aaron Carass, and Jacob Reinhold from Johns Hopkins University for insightful conversations about determinism in GPUs and floating point precision. This material is based upon work supported by the National Science Foundation Graduate Research Fellowship under Grant No. DGE-1746891. Support for this work included funding from the Intramural Research Program of the NIH Clinical Center and the Department of Defense in the Center for Neuroscience and Regenerative Medicine, the National Multiple Sclerosis Society RG-1507-05243 (Pham), and NIH grant 1R01EB017230-01A1 (Landman), as well as NSF 1452485 (Landman). This work received support from the Advanced Computing Center for Research and Education (ACCRE) at the Vanderbilt University, Nashville, TN, as well as in part by ViSE/VICTR VR3029. We also extend gratitude to NVIDIA for their support by means of the NVIDIA hardware grant.

References

1. Act, A.: Health insurance portability and accountability act of 1996. Public law **104**, 191 (1996)
2. Alistarh, D., Grubic, D., Li, J., Tomioka, R., Vojnovic, M.: QSGD: Communication-efficient SGD via gradient quantization and encoding. In: Guyon, I., et al. (eds.) Advances in Neural Information Processing Systems, vol. 30, pp. 1709–1720. Curran Associates, Inc. (2017)
3. Bengio, Y.: Practical recommendations for gradient-based training of deep architectures. In: Montavon, G., Orr, G.B., Müller, K.-R. (eds.) Neural Networks: Tricks of the Trade. LNCS, vol. 7700, pp. 437–478. Springer, Heidelberg (2012). https://doi.org/10.1007/978-3-642-35289-8_26
4. Chang, K., et al.: Distributed deep learning networks among institutions for medical imaging. J. Am. Med. Inform. Assoc. **25**(8), 945–954 (2018)
5. Fetzer, D.T., West, O.C.: The HIPAA privacy rule and protected health information: implications in research involving DICOM image databases. Acad. Radiol. **15**(3), 390–395 (2008)
6. French, R.M.: Catastrophic forgetting in connectionist networks. Trends Cogn. Sci. **3**(4), 128–135 (1999)
7. Gibson, E., et al.: Inter-site variability in prostate segmentation accuracy using deep learning. In: Frangi, A.F., Schnabel, J.A., Davatzikos, C., Alberola-López, C., Fichtinger, G. (eds.) MICCAI 2018. LNCS, vol. 11073, pp. 506–514. Springer, Cham (2018). https://doi.org/10.1007/978-3-030-00937-3_58
8. Goodfellow, I., Bengio, Y., Courville, A.: Deep Learning. MIT Press, Cambridge (2016)
9. Hitaj, B., Ateniese, G., Perez-Cruz, F.: Deep models under the GAN: information leakage from collaborative deep learning. In: Proceedings of the 2017 ACM SIGSAC Conference on Computer and Communications Security, pp. 603–618 (2017)
10. Kingma, D.P., Ba, J.: Adam: a method for stochastic optimization. arXiv preprint arXiv:1412.6980 (2014)
11. Kirkpatrick, J., et al.: Overcoming catastrophic forgetting in neural networks. Proc. Nat. Acad. Sci. **114**(13), 3521–3526 (2017)
12. Konecnỳ, J., McMahan, H.B., Ramage, D., Richtárik, P.: Federated optimization: Distributed machine learning for on-device intelligence. arXiv preprint arXiv:1610.02527 (2016)
13. LeCun, Y., Cortes, C., Burges, C.: MNIST handwritten digit database. ATT Labs (2010). http://yann.lecun.com/exdb/mnist

14. Li, W., et al.: Privacy-preserving federated brain tumour segmentation. In: Suk, H.-I., Liu, M., Yan, P., Lian, C. (eds.) MLMI 2019. LNCS, vol. 11861, pp. 133–141. Springer, Cham (2019). https://doi.org/10.1007/978-3-030-32692-0_16

15. Li, X., Huang, K., Yang, W., Wang, S., Zhang, Z.: On the convergence of FedAvg on Non-IID data. arXiv preprint arXiv:1907.02189 (2019)

16. Lin, Y., Han, S., Mao, H., Wang, Y., Dally, W.J.: Deep gradient compression: reducing the communication bandwidth for distributed training. arXiv preprint arXiv:1712.01887 (2017)

17. Luxton, D.D., Kayl, R.A., Mishkind, M.C.: mhealth data security: the need for HIPAA-compliant standardization. Telemedicine and e-Health 18(4), 284–288 (2012)

18. McClure, P., et al.: Distributed weight consolidation: a brain segmentation case study. In: Advances in Neural Information Processing Systems, pp. 4093–4103 (2018)

19. McMahan, H.B., Moore, E., Ramage, D., Hampson, S., et al.: Communication-efficient learning of deep networks from decentralized data. arXiv preprint arXiv:1602.05629 (2016)

20. Nguyen, C.V., Li, Y., Bui, T.D., Turner, R.E.: Variational continual learning. arXiv preprint arXiv:1710.10628 (2017)

21. NVIDIA: Tensorflow determinism (2020). https://github.com/NVIDIA/framework-determinism. Accessed 27 Jun 2020

22. Pan, S.J., Yang, Q., et al.: A survey on transfer learning. IEEE Trans. Knowl. Data Eng. 22(10), 1345–1359 (2010)

23. MONAI (2020). https://monai.io/. Accessed 14 Jul 2020

24. Remedios, S.W.: Federated gradient averaging implementation (2020). https://github.com/sremedios/federated_gradient_averaging. Accessed 27 Jun 2020

25. Remedios, S.W., et al.: Distributed deep learning across multisite datasets for generalized CT hemorrhage segmentation. Med. Phys. 47(1), 89–98 (2020)

26. Ronneberger, O., Fischer, P., Brox, T.: U-Net: convolutional networks for biomedical image segmentation. In: Navab, N., Hornegger, J., Wells, W.M., Frangi, A.F. (eds.) MICCAI 2015. LNCS, vol. 9351, pp. 234–241. Springer, Cham (2015). https://doi.org/10.1007/978-3-319-24574-4_28

27. Ruder, S.: An overview of gradient descent optimization algorithms. arXiv preprint arXiv:1609.04747 (2016)

28. Sheller, M.J., et al.: Federated learning in medicine: facilitating multi-institutional collaborations without sharing patient data. Scientific reports 10(1), 1–12 (2020)

29. Shokri, R., Shmatikov, V.: Privacy-preserving deep learning. In: Proceedings of the 22nd ACM SIGSAC Conference on Computer and Communications Security, pp. 1310–1321 (2015)

30. Shokri, R., Stronati, M., Song, C., Shmatikov, V.: Membership inference attacks against machine learning models. In: 2017 IEEE Symposium on Security and Privacy (SP), pp. 3–18. IEEE (2017)

31. Thompson, L.A., Black, E., Duff, W.P., Black, N.P., Saliba, H., Dawson, K.: Protected health information on social networking sites: ethical and legal considerations. J. Med. Internet Res. 13(1), e8 (2011)

32. Wen, W., et al.: TernGrad: ternary gradients to reduce communication in distributed deep learning. In: Advances in Neural Information Processing Systems, pp. 1509–1519 (2017)

33. Ye, M., Abbe, E.: Communication-computation efficient gradient coding. arXiv preprint arXiv:1802.03475 (2018)

34. Yosinski, J., Clune, J., Bengio, Y., Lipson, H.: How transferable are features in deep neural networks? In: Advances in neural information processing systems, pp. 3320–3328 (2014)
35. Yu, H., Yang, S., Zhu, S.: Parallel restarted SGD with faster convergence and less communication: demystifying why model averaging works for deep learning. In: Proceedings of the AAAI Conference on Artificial Intelligence, vol. 33, pp. 5693–5700 (2019)
36. Zhang, S., Zhang, C., You, Z., Zheng, R., Xu, B.: Asynchronous stochastic gradient descent for DNN training. In: 2013 IEEE International Conference on Acoustics, Speech and Signal Processing (ICASSP), pp. 6660–6663. IEEE (2013)

Federated Learning for Breast Density Classification: A Real-World Implementation

Holger R. Roth[1]([✉]), Ken Chang[2], Praveer Singh[2], Nir Neumark[2], Wenqi Li[1], Vikash Gupta[3], Sharut Gupta[2], Liangqiong Qu[4], Alvin Ihsani[1], Bernardo C. Bizzo[2], Yuhong Wen[1], Varun Buch[2], Meesam Shah[5], Felipe Kitamura[6], Matheus Mendonça[6], Vitor Lavor[6], Ahmed Harouni[1], Colin Compas[1], Jesse Tetreault[1], Prerna Dogra[1], Yan Cheng[1], Selnur Erdal[3], Richard White[3], Behrooz Hashemian[2], Thomas Schultz[2], Miao Zhang[4], Adam McCarthy[2], B. Min Yun[2], Elshaimaa Sharaf[2], Katharina V. Hoebel[7], Jay B. Patel[7], Bryan Chen[7], Sean Ko[7], Evan Leibovitz[2], Etta D. Pisano[2], Laura Coombs[5], Daguang Xu[1], Keith J. Dreyer[2], Ittai Dayan[2], Ram C. Naidu[2], Mona Flores[1], Daniel Rubin[4], and Jayashree Kalpathy-Cramer[2]

[1] NVIDIA, Santa Clara, USA
hroth@nvidia.com
[2] Massachusetts General Hospital, Boston, USA
[3] Mayo Clinic, Jacksonville, USA
[4] Stanford University, Stanford, USA
[5] American College of Radiology, Reston, USA
[6] Diagnósticos da América (DASA), São Paulo, Brazil
[7] Massachusetts Institute of Technology, Cambridge, USA

Abstract. Building robust deep learning-based models requires large quantities of diverse training data. In this study, we investigate the use of federated learning (FL) to build medical imaging classification models in a real-world collaborative setting. Seven clinical institutions from across the world joined this FL effort to train a model for breast density classification based on Breast Imaging, Reporting & Data System (BI-RADS). We show that despite substantial differences among the datasets from all sites (mammography system, class distribution, and data set size) and without centralizing data, we can successfully train AI models in federation. The results show that models trained using FL perform 6.3% on average better than their counterparts trained on an institute's local data alone. Furthermore, we show a 45.8% relative improvement in the models' generalizability when evaluated on the other participating sites' testing data.

Keywords: Federated learning · Breast density classification · BI-RADS · Mammography

© Springer Nature Switzerland AG 2020
S. Albarqouni et al. (Eds.): DART 2020/DCL 2020, LNCS 12444, pp. 181–191, 2020.
https://doi.org/10.1007/978-3-030-60548-3_18

1   Introduction

Advancements in medical image analysis over the last several years have been dominated by deep learning (DL) approaches. However, it is well known that DL requires large quantities of data to train robust and clinically useful models [5,6]. Often, hospitals and other medical institutes need to collaborate and host centralized databases for the development of clinically useful models. This overhead can quickly become a logistical challenge and usually requires a time-consuming approval process due to data privacy and ethical concerns associated with data sharing in healthcare [12]. Even when these challenges can be addressed, data is valuable, and institutions may prefer not to share full datasets. Furthermore, medical data can be large, and it may be prohibitively expensive to acquire storage for central hosting [4]. One approach to combat the data sharing hurdles is federated learning (FL) [16], where only model weights are shared between participating institutions without sharing the raw data.

To investigate the performance of FL in the real world, we conducted a study to develop a breast density classification model using mammography data. An international group of hospitals and medical imaging centers joined this collaborative effort to train the model in purely data-decentralized fashion without needing to share any data. This is in contrast to previous studies in which the FL environment was only simulated [14,21]. We do not have centralized training experiments as references before starting the FL tasks, which places higher requirements on the robustness of the algorithms and selection of hyperparameters.

1.1   Related Works

Breast Density Scoring: The classification of breast density is quintessential for breast imaging to estimate the extent of fibroglandular tissue related to the patient's risk of developing breast cancer [2,19]. Women with a high mammographic breast density (>75%) have a four- to five-fold increase in risk for breast cancer compared to women having a lower breast density [3,26]. This condition affects roughly half of American women between the ages of 40 to 74 [7,25]. Patients identified with dense breast tissue may have masked tumors and benefit from supplemental imaging such as MRI or ultrasound [13]. High mammographic breast density impairs the sensitivity and specificity of breast cancer screening, possibly because (small) malignant lesions are not detectable even when they are present [17]. The standard evaluation metric for reporting breast density is the Breast Imaging Reporting and Data System (BI-RADS), based on 2D mammography [22]. Scans are categorized into one of four classes: (a) fatty, (b) scattered, (c) heterogeneously dense, and (d) extremely dense.

Due to the subjective nature of the BI-RADS criteria, there can be substantial inter-rater variability between pairs of clinicians. Sprague et al. [24] found that the likelihood of a mammogram being read as dense varies from radiologist to radiologist between 6.3% to 84.5%. Ooms et al. find that the overall agreement between four observers (inter-rater agreement) in terms of the overall weighted kappa was 0.77 [17]. Another study reported the inter-rater variability to be simple kappa = 0.58 among 34 community radiologists [23]. Even the intra-rater agreement in the assessment of BI-RADS breast density can be relatively low. Spayne et al. [23] showed that the intra-rater agreement was below 80% when evaluating the same mammography exam within a 3- to 24- month period. Recent work on applying DL for mammography breast density classification [13] achieved a linear kappa of 0.67 when comparing the DL model's predictions to the assessments of the original interpreting radiologist.

Federated Learning: Federated learning has recently been described as being instrumental for the future of digital health [20]. FL enables collaborative and decentralized DL training without sharing any raw patient data [16]. Each client in FL trains locally on their data and then submits their model parameters to a server that accumulates and aggregates the model updates from each client. Once a certain number of clients have submitted their updates, the aggregated model parameters are redistributed to the clients, and a new round of local training starts. While out of the scope of this work, FL can also be combined with additional privacy-preserving measures to avoid potential reconstruction of training data through model inversion if the model parameters would be exposed to an adversary [14]. Recent works have shown the applicability of FL to medical imaging tasks [14,21]. The security and privacy-preserving aspects of federated machine learning in medical imaging have been discussed in more detail by Kaissis et al. [9].

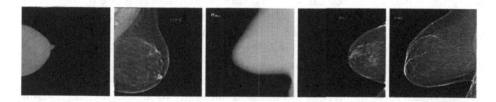

Fig. 1. Mammography data examples from different sites after resizing the original images to a resolution of 224 × 224. No special normalization was applied in order to keep the scanners' original intensity distribution that can be observed in 4.

2   Method

We implemented our FL approach in a real-world setting with participation from seven international clients.

Datasets: The mammography data was retrospectively selected after Institutional Review Board (IRB) approval as part of standard mammography screening protocols. The BI-RADS breast density class from the original interpreting radiologist was collected from the reports available in the participating hospitals' medical records and includes images from digital screening mammography

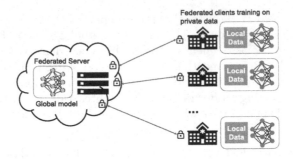

Fig. 2. Federated learning in medical imaging. The central server communicates with clients from multi-national institutions without exchanging any sensitive raw data. Still, the global model benefits from weights and gradients from clients' local models to achieve higher overall performance.

(Fig. 1). Clients 1 to 3 utilized the multi-institutional dataset previously described in [18], which was split by the digital mammography system used to acquire the image to account for different dataset sources.

Each client's data exhibited their own characteristics of detector type, image resolution, and mammography type. Furthermore, the number of training images varied significantly among clients, as shown in Table 1. The distributions of the different BI-RADS categories were markedly different at some clients but generally followed the distribution known from the literature, with more images in the categories b and c [18], see Fig. 3. Given these differences that are quite typical for real-world multi-institutional datasets, we can see that the data used in this study is non-independent and identically distributed (non-IID).

Intensity distributions among different sites also varied markedly, as can be observed in Fig. 4. This variance is due to the differences in imaging protocols and digital mammography systems used at each data contributing site. No attempt to consolidate these differences was made in our study to investigate the domain shift challenges proposed by this non-IID data distribution.

Table 1. Dataset characteristics at each client. Image resolution is shown in megapixels (MP).

Institution	Image resolution	Detector type	Image type	Bits	# Train	# Val.	# Test
client1	23.04	Direct	2D	12	22933	3366	6534
client2	.02 to 4.39	Direct	2D	12	8365	1216	2568
client3	4.39 to 13.63	Direct	2D	14	44115	6336	12676
client4	4 to 28	Direct/Scintillator	2D	12	7219	1030	2069
client5	8.48 to 13.63	Direct	2D	12	6023	983	1822
client6	8.6 to 13.63	Direct	2D	12	6874	853	1727
client7	1 to 136	Direct/Scintillator	2D/tomosynthesis	10/12	4021	664	1288

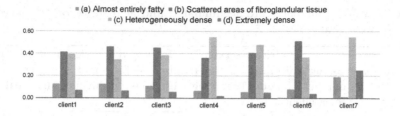

Fig. 3. Class distribution at different client sites as a fraction of their total data.

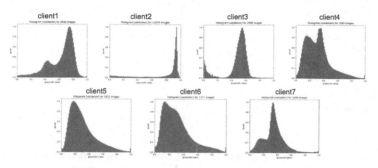

Fig. 4. Intensity distribution at different sites.

Client-Server-Based Federated Learning: In its typical form, FL utilizes a client-server setup (Fig. 2). Each client trains the same model architecture locally on their data. Once a certain number of clients finishes a round of local training, the updated model weights (or their gradients) are sent to the server for aggregation. After aggregation, the new weights on the server are re-distributed to the clients, and the next round of local model training begins. After a certain number of FL rounds, the models at each client converge. Each client is allowed to select their locally best model by monitoring a certain performance metric on a local hold-out validation set. The client can select either the global model returning from the server after averaging or any intermediate model considered best during local training based on their validation metric. In our experiments, we implement the `FederatedAveraging` algorithm proposed in [16]. While there exist variations of this algorithm to address particular learning tasks, in its most general form, FL tries to minimize a global loss function $\mathcal{L}$ which can be a weighted combination of K local losses $\{\mathcal{L}_k\}_{k=1}^{K}$, each of which is computed on a client k's local data. Hence, FL can be formulated as the task of finding the model parameters ϕ that minimize L given some local data $X_k \in X$, where X would be the combination of all local datasets.

$$\min_{\phi} \mathcal{L}(X;\phi) \quad \text{with} \quad \mathcal{L}(X;\phi) = \sum_{k=1}^{K} w_k \, \mathcal{L}_k(X_k;\phi), \tag{1}$$

where $w_k > 0$ denotes the weight coefficients for each client k, respectively. Note that the local data X_k is never shared among the different clients. Only the model weight differences are accumulated and aggregated on the server as shown in Algorithm 1.

Algorithm 1 Client-server federated learning with FederatedAveraging [16,14]. T is the number of federated learning rounds and n_k is the number of LocalTraining iterations minimizing the local loss $\mathcal{L}_k(X_k; \phi^{(t-1)})$ for a client k.

1: **procedure** FEDERATED LEARNING
2:    Initialize weights: $\phi^{(0)}$
3:    **for** $t \leftarrow 1 \cdots T$ **do**
4:        **for** *client* $k \leftarrow 1 \cdots K$ **do**   ▷ *Executed in parallel*
5:            Send $\phi^{(t-1)}$ to client k
6:            Receive $(\Delta\phi_k^{(t)}, n_k)$ from client's LocalTraining($\phi^{(t-1)}$)
7:        **end for**
8:        $\phi_k^{(t)} \leftarrow \phi^{(t-1)} + \Delta\phi_k^{(t)}$
9:        $\phi^{(t)} \leftarrow \frac{1}{\sum_k n_k} \sum_k (n_k \cdot \phi_k^{(t)})$
10:   **end for**
11:   **return** $\phi^{(t)}$
12: **end procedure**

In this work, we choose a softmax cross-entropy loss which is commonly used for multi-class classification tasks: $\mathcal{L}_0 = -\sum_{i=1}^{C} y_i \log(p_i)$; with $C = 4$ being the number of classes. Here, p_i is the predicted probability for a class i from the final softmax activated output layer of our neural network $f(x)$ and y is the one-hot encoded ground truth label for a given image.

Classification Model and Implementation: In this work, we do not focus on developing a new model architecture but instead focus on showing how FL works in a real-world collaborative training situation. We implement a DenseNet-121 [8] model as a backbone and append a fully-connected layer with four outputs to its last feature layer to classify a mammography image as one of the four BI-RADS categories. The FL framework is implemented in Tensorflow[1] and utilizes the NVIDIA Clara Train SDK[2] to enable the communication between server and clients as well as to standardize the training configuration among clients. Each client employed an NVIDIA GPU with at least 12 GB memory.

All mammography images were normalized to an intensity range of $[0\ldots 1]$ and resampled to a resolution of 224×224. We include both left and right breast images and all available views (craniocaudal and mediolateral oblique) in training. Each client separated their dataset into training, validation, and testing sets on the patient level (see Table 1). At inference time, predictions from all images from a given patient were averaged together to give a patient-level prediction.

Each client trained for one epoch before sending their updated model weights to the server for aggregation, and the server waited for all clients before performing a weighted sum of the clients' weight differences. We used initial learning rates of 1e−4 with step-based learning rate decay, Adam optimization for each

---

[1] https://www.tensorflow.org/.
[2] https://developer.nvidia.com/clara.

client, and model weight decay. A mini-batch of size 32 was sampled from the dataset such that all categories were equally represented during training. Random spatial flips, rotations between ±45°, and intensity shifts were used as on-the-fly image augmentation to avoid overfitting to the training data. The FL training was run for 300 rounds of local training and weight aggregations, which took about 36 h. After the FL training is finished, each client's best local model is shared with all other clients and tested on their test data to evaluate the models' generalizability.

In an additional experiment, we use the locally best models each client receives after FL to execute a second round of local fine-tuning based on this model. This additional "adaptation" step can improve a client's model on their local data.

Evaluation Metric: We utilize Cohen's linear weighed kappa[3] to evaluate the locally best models' performance before and after federated learning in comparison with the radiologists' ground truth assessments. The kappa score is a number between −1 and 1. Scores above 0.8 are generally considered very good agreement, while zero or lower would mean no agreement (practically random assignment of labels). A kappa of 0.21 to 0.40, 0.41 to 0.60, and 0.61 to 0.80 represents fair, moderate, and substantial agreement, respectively [11]. The kappa measure has been chosen to be directly comparable to previous literature on breast density classification in mammography [13,17,23].

3   Results

In Table 2, we show the performance of locally best models (selected by best validation score on local data) using local training data alone as well as after federated learning. On average, a 6.3% relative improvement can be observed when the model is applied to a client's test data (diag. mean). We also observe a general improvement of these best local models applied to the different clients' test data. Here, the generalizability (off-diag. mean) of the models improved by 45.8% on average.

Figure 5 summarizes the kappa scores for local training, after FL, including after local fine-tuning, which improves a given model's performance on the client's local test data in all but one client.

---

[3] https://scikit-learn.org/stable/modules/generated/sklearn.metrics.cohen_kappa_sco re.html.

Table 2. Performance of locally best models (selected by best validation score on local data) using (a) local training data alone and (b) after federated learning.

(a) Local:				Test					(b) Federated:				Test				
client	1	2	3	4	5	6	7		client	1	2	3	4	5	6	7
1	0.62	0.59	0.44	0.02	0.02	-0.01	0.04		1	0.62	0.62	0.48	0.15	0.23	0.24	0.11
2	0.15	0.56	0.02	-0.01	-0.00	0.00	-0.01		2	0.22	0.65	0.11	0.04	0.00	0.00	-0.01
3	0.19	0.01	0.64	0.02	0.07	0.00	0.05		3	0.41	0.17	0.63	0.07	-0.00	0.01	-0.01
4	0.11	0.02	-0.00	0.63	0.52	0.61	0.50		4	0.06	0.48	-0.02	0.69	0.57	0.65	0.52
5	-0.00	-0.01	-0.03	0.54	0.62	0.65	0.31		5	0.24	0.13	0.02	0.64	0.62	0.69	0.52
6	0.01	0.11	-0.02	0.49	0.59	0.71	0.32		6	0.23	0.01	-0.00	0.53	0.68	0.76	0.31
7	0.03	0.05	-0.05	0.40	0.37	0.46	0.69		7	0.10	0.21	0.13	0.55	0.44	0.52	0.77
									Global	0.51	0.52	0.49	0.31	0.4852	0.31	0.0893

diag. mean			0.64	diag. mean		0.68
off-diag. mean			0.18	off-diag. mean		0.26

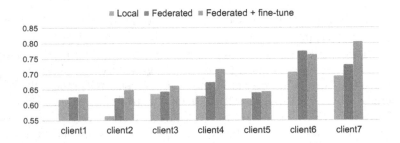

- Local ■ Federated ■ Federated + fine-tune

Fig. 5. Weighted linear kappa performance before and after federated learning, and after an additional round of local fine-tuning at each local site.

4   Discussion and Conclusions

Given our experimental results, we can see that federated learning (FL) in a real-world scenario can both achieve more accurate models locally as well as increase the generalizability of these models to data from other sources, such as test data from other clients. This improvement is due to the effectively larger training set made available through FL without the need to share any data directly. While we cannot directly compare to a centralized training setting due to the nature of performing FL in a real-world setting, we observed that the average performance of models is similar to values reported in the literature on centralized datasets. For example, Lehman et al. [13] reported a linear kappa value of 0.67 when applying DL for mammography breast density classification. We achieved an average performance of local models of 0.68 in the FL setting, confirming the ability of FL to achieve models comparable to models trained when the data is accumulated in a central database. However, while the generalizability is improved, it is still not comparable to the performance on local test sets. In particular, the final global model is not near any acceptable performance on any of the local test datasets. The heterogeneity in results across institutions illustrates the difficulties in training models that are generalizable. In practice, some local adaptation (fine-tuning, see Fig. 5) or at least model selection based on local validation data (see diagonal of Table 2) is needed.

In this work, we deliberately did not attempt any data harmonization methods to study the effect of different data domains. The marked differences in intensity distributions due to different mammography systems are observable in Fig. 4. Future work might explore the use of histogram equalization and other techniques [1,10] to harmonize non-IID data across different sites or investigate built-in strategies for domain adaptation within the FL framework [15]. Similarly, we did not fully address issues of data size heterogeneity and class imbalance within our FL framework. For example, client 7 had almost no category (b) samples due to their local labeling practices required by their clinical protocol. Future work could incorporate training strategies such as client-specific local training iterations, other mini-batch sampling strategies, and loss functions. We also did not attempt privacy-preservation techniques that would reduce the chance of model inversion and potential data leakage based on the trained models. Differential privacy could easily be applied to our framework, and it has been shown that it can achieve comparable results to the vanilla FL setting [14].

Despite these challenges, we were able to train mammography models in a real-world FL setting that improved the performance of locally trained models alone, illustrating the promise of FL for building clinically-applicable models and sidestepping the need for accumulating a centralized dataset.

Acknowledgements. Research reported in this publication was supported by a training grant from the National Institute of Biomedical Imaging and Bioengineering (NIBIB) of the National Institutes of Health (NIH) under award number 5T32EB1680 to K. Chang and J. B. Patel and by the National Cancer Institute (NCI) of the NIH under Award Number F30CA239407 to K. Chang. This study was supported by NIH grants U01CA154601, U24CA180927, U24CA180918, and U01CA242879, and National Science Foundation (NSF) grant NSF1622542 to J. Kalpathy-Cramer.

References

1. Baweja, C., Glocker, B., Kamnitsas, K.: Towards continual learning in medical imaging. arXiv preprint arXiv:1811.02496 (2018)
2. Boyd, N., et al.: Quantitative classification of mammographic densities and breast cancer risk: results from the Canadian national breast screening study. JNCI J. Nat. Cancer Inst. **87**(9), 670–675 (1995)
3. Boyd, N.F., et al.: Mammographic density and the risk and detection of breast cancer. N. Engl. J. Med. **356**(3), 227–236 (2007)
4. Chang, K., et al.: Distributed deep learning networks among institutions for medical imaging. J. Am. Med. Inform. Assoc. **25**(8), 945–954 (2018)
5. Chang, K., et al.: Multi-institutional assessment and crowdsourcing evaluation of deep learning for automated classification of breast density. J. Am. Coll. Radiol. (2020). https://doi.org/10.1016/j.jacr.2020.05.015
6. Dunnmon, J.A., Yi, D., Langlotz, C.P., Ré, C., Rubin, D.L., Lungren, M.P.: Assessment of convolutional neural networks for automated classification of chest radiographs. Radiology **290**(2), 537–544 (2019)
7. Ho, J.M., Jafferjee, N., Covarrubias, G.M., Ghesani, M., Handler, B.: Dense breasts: a review of reporting legislation and available supplemental screening options. AJR Am. J. Roentgenol. **203**(2), 449–456 (2014)

8. Huang, G., Liu, Z., Van Der Maaten, L., Weinberger, K.Q.: Densely connected convolutional networks. In: Proceedings of the IEEE Conference on Computer Vision and Pattern Recognition, pp. 4700–4708 (2017)
9. Kaissis, G.A., Makowski, M.R., Rückert, D., Braren, R.F.: Secure, privacy-preserving and federated machine learning in medical imaging. Nat. Mach. Intell. **2**, 1–7 (2020)
10. Karani, N., Chaitanya, K., Baumgartner, C., Konukoglu, E.: A lifelong learning approach to brain MR segmentation across scanners and protocols. In: Frangi, A.F., Schnabel, J.A., Davatzikos, C., Alberola-López, C., Fichtinger, G. (eds.) MICCAI 2018. LNCS, vol. 11070, pp. 476–484. Springer, Cham (2018). https://doi.org/10.1007/978-3-030-00928-1_54
11. Landis, J.R., Koch, G.G.: The measurement of observer agreement for categorical data. Biometrics **33**, 159–174 (1977)
12. Larson, D.B., Magnus, D.C., Lungren, M.P., Shah, N.H., Langlotz, C.P.: Ethics of using and sharing clinical imaging data for artificial intelligence: a proposed framework. Radiology **295**, 192536 (2020)
13. Lehman, C.D., et al.: Mammographic breast density assessment using deep learning: clinical implementation. Radiology **290**(1), 52–58 (2019)
14. Li, W., et al.: Privacy-preserving federated brain tumour segmentation. In: Suk, H.-I., Liu, M., Yan, P., Lian, C. (eds.) MLMI 2019. LNCS, vol. 11861, pp. 133–141. Springer, Cham (2019). https://doi.org/10.1007/978-3-030-32692-0_16
15. Li, X., Gu, Y., Dvornek, N., Staib, L., Ventola, P., Duncan, J.S.: Multi-site FMRI analysis using privacy-preserving federated learning and domain adaptation: Abide results. arXiv preprint arXiv:2001.05647 (2020)
16. McMahan, B., Moore, E., Ramage, D., Hampson, S., y Arcas, B.A.: Communication-efficient learning of deep networks from decentralized data. In: Artificial Intelligence and Statistics, pp. 1273–1282 (2017)
17. Ooms, E., et al.: Mammography: interobserver variability in breast density assessment. Breast **16**(6), 568–576 (2007)
18. Pisano, E.D., et al.: Diagnostic performance of digital versus film mammography for breast-cancer screening. N. Engl. J. Med. **353**(17), 1773–1783 (2005)
19. Razzaghi, H., Troester, M.A., Gierach, G.L., Olshan, A.F., Yankaskas, B.C., Millikan, R.C.: Mammographic density and breast cancer risk in white and African American women. Breast Cancer Res. Treat. **135**(2), 571–580 (2012)
20. Rieke, N., et al.: The future of digital health with federated learning. arXiv preprint arXiv:2003.08119 (2020)
21. Sheller, M.J., Reina, G.A., Edwards, B., Martin, J., Bakas, S.: Multi-institutional deep learning modeling without sharing patient data: a feasibility study on brain tumor segmentation. In: Crimi, A., Bakas, S., Kuijf, H., Keyvan, F., Reyes, M., van Walsum, T. (eds.) BrainLes 2018. LNCS, vol. 11383, pp. 92–104. Springer, Cham (2019). https://doi.org/10.1007/978-3-030-11723-8_9
22. Sickles, E., d'Orsi, C., Bassett, L., Appleton, C., Berg, W., Burnside, E., et al.: ACR BI-RADS® mammography. In: ACR BI-RADS® Atlas, Breast Imaging Reporting and Data System, vol. 5, p. 2013 (2013)
23. Spayne, M.C., Gard, C.C., Skelly, J., Miglioretti, D.L., Vacek, P.M., Geller, B.M.: Reproducibility of bi-rads breast density measures among community radiologists: a prospective cohort study. Breast J. **18**(4), 326–333 (2012)
24. Sprague, B.L., et al.: Variation in mammographic breast density assessments among radiologists in clinical practice: a multicenter observational study. Ann. Intern. Med. **165**(7), 457–464 (2016)

25. Sprague, B.L., et al.: Prevalence of mammographically dense breasts in the United States. JNCI J. Nat. Cancer Inst. **106**(10), dju255 (2014)
26. Yaghjyan, L., et al.: Mammographic breast density and subsequent risk of breast cancer in postmenopausal women according to tumor characteristics. J. Nat. Cancer Inst. **103**(15), 1179–1189 (2011)

Automated Pancreas Segmentation Using Multi-institutional Collaborative Deep Learning

Pochuan Wang[1], Chen Shen[2], Holger R. Roth[3], Dong Yang[3], Daguang Xu[3], Masahiro Oda[2], Kazunari Misawa[4], Po-Ting Chen[5], Kao-Lang Liu[5], Wei-Chih Liao[5], Weichung Wang[1], and Kensaku Mori[2(✉)]

[1] National Taiwan University, Taipei, Taiwan
[2] Nagoya University, Nagoya, Japan
kensaku@is.nagoya-u.ac.jp
[3] NVIDIA Corporation, Santa Clara, USA
[4] Aichi Cancer Center, Nagoya, Japan
[5] National Taiwan University Hospital, Taipei, Taiwan

Abstract. The performance of deep learning based methods strongly relies on the number of datasets used for training. Many efforts have been made to increase the data in the medical image analysis field. However, unlike photography images, it is hard to generate centralized databases to collect medical images because of numerous technical, legal, and privacy issues. In this work, we study the use of federated learning between two institutions in a real-world setting to collaboratively train a model without sharing the raw data across national boundaries. We quantitatively compare the segmentation models obtained with federated learning and local training alone. Our experimental results show that federated learning models have higher generalizability than standalone training.

Keywords: Federated learning · Pancreas segmentation · Neural architecture search

1 Introduction

Recently, deep neural networks (DNNs) based methods have been widely utilized for medical imaging research. High-performing models that are clinically useful always require vast, varied, and high-quality datasets. However, it is expensive to collect a large number of datasets, especially in the medical field. Only well-trained experts can generate acceptable annotations for DNN training, making annotated medical images even more scarce. Furthermore, medical images from a single institution can be biased towards specific pathologies, equipment, acquisition protocols, and patient populations. The low generalizability of DNNs models

---

P. Wang and C. Shen—Equal contribution.

trained on insufficient datasets is critical when applying deep learning methods for clinical usages.

To improve the robustness with scant data, fine-tuning is an alternative way to learn the knowledge from pre-trained DNNs. The fine-tuning technique starts training from a pre-trained weight instead of random initialization, which has been proved helpful in medical image analysis [7,8], which exceeds the performance on training a DNN from scratch. However, fine-tuned models can still have high deficiencies in generalizability [1]. When a model is pre-trained on one data (source data) and then fine-tuned on another data (target data), the trained model tends to fit on target data but lose the representation on source data [3].

Federated learning (FL) [4] is an innovation for solving this issue. It can collaboratively train the DNNs using the datasets from multiple institutions without creating a centralized dataset [2,6]. Each institution (client) trains with local data using the same network architecture decided in advance. After a certain amount of local training, each institution regularly sends the trained model to the server. The server only centralizes the weights of the model to aggregate, and then send them back to each client.

In this work, we collaboratively generated and evaluated an FL model for pancreas segmentation without sharing the data. Our data consists of healthy and unhealthy pancreas collected at the two institutions from different countries (Taiwan and Japan). Throughout this study, we utilized the model from coarse-to-fine network architecture search (C2FNAS) [10] with an additional variational auto-encoder (VAE) [5] branch to the encoder endpoint. FL dramatically improved the generalizability of models on server-side and client-side for both datasets. To the best of our knowledge, this is the first time performing FL for building a pancreas segmentation model from data hosted at multi-national sites.

2 Methods

2.1 Federated Learning

FL can be categorized into different types based on the distribution characteristics of data [9]. In this work, we only focus on horizontal architecture, which is illustrated in Fig. 1. This type of FL allows us to train with datasets from different samples distributed across clients.

A horizontal FL system consists of two parts: *server* and *clients*. The server manages the training process and generates a *global model*, and the client train with local data to produces a *local model*. The server receives trained weights from each client and aggregates them into a global model. The clients train with the local dataset and send the weights to the server. We call the process of generating one global model one round.

The workflow consists of the following steps:

1. Start the server. The server-side sets the gPRC communication ports, SSL certificate, the maximum and minimum numbers of clients.

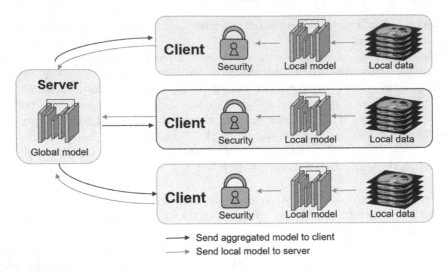

Fig. 1. The architecture of federated learning system.

2. Start the client. Use client-side configuration to initialize. Then use the credential to make a login request to the server.
3. Client-side downloads the current global model from the server and fine-tuning the model with the local dataset. Then, only submit the model to the server and wait for other clients.
4. Once the server receives the model from a previously defined minimum number of clients, it will aggregate them into a new global model.
5. The server updates the global model and finishes one round.
6. Go back to 3. for another round.

The model shared among the server and clients is only weight parameters, protecting the privacy for local data. To build the server-client trust, the server-side uses token throughout the process. SSL certificate authority and gPRC communication ports were adopted to improve security.

2.2 Data Collection

We use two physically separated clients in this work in order to try FL in the real-world setting. Two different datasets from two institutions from two different countries were applied.

For Client 1, we utilize 420 portal-venous phase abdominal CT images collected for preoperative planning in gastric surgery, so the stomach part is inflated. For the pancreas, we did not notice any particular abnormalities. The resolution of volumes are (0.58–0.98, 0.58–0.98, 0.16–1.0) in the voxel spacing (x, y, z) in millimeter. Only pancreas regions are manually annotated using semi-automated segmentation tools. We randomly split the data set into 252 training volumes, 84 validation volumes, and 84 testing volumes.

For Client 2's dataset, we collected 486 contrast-enhanced abdominal CT images, where all volumes are from patients with pancreatic cancer. Among the whole dataset, the voxel spacing (x, y, z) in millimeter of 40 volumes are (0.68, 0.68, 1.0) and the rest 446 volume are (0.68, 0.68, 5.0) in millimeter. The segmentation labels contain the normal part of the pancreas and the tumor of pancreatic cancer. All the labels are manually segmented by physicians. We split client 2's dataset into training, validation and testing sets randomly, the training set contains 286 volumes, the validation set contains 100 volumes and the testing set also contains 100 volumes.

2.3   Data Pre-processing

We re-sample the resolution of all volumes to isotropic spacing 1.0 mm × 1.0 mm × 1.0 mm, and apply intensity clipping with minimum Hounsfield unit (HU) intensity −200 and maximum intensity 250. Then we re-scale the value range to $[-1.0, 1.0]$.

2.4   Neural Network Model

We utilized the resulting model of coarse-to-fine network architecture search (C2FNAS) [10]. The C2FNAS search algorithm performs a coarse-level search followed by a fine-level search to determine the optimal neural network architecture for 3D medical image segmentation. In the coarse-level search, C2FNAS searched for the topology of U-Net like models. In the fine-level search, C2FNAS searched for the optimal operations (including 2D convolution, 3D convolution, and pseudo-3D convolution) for each module from the previous search results.

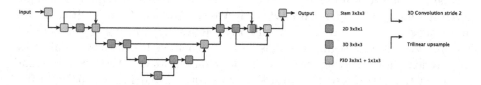

Fig. 2. Model architecture of C2FNAS

We add a VAE branch to the encoder endpoint of the C2FNAS model. The VAE branch shares encoder layers with C2FNAS and estimates the mean and standard deviation of encoded features for input image reconstruction. Two further losses, L_{KL} and L_2, are introduced in [5] are required for the VAE branch. L_{KL} estimates the distance of mean vector and standard deviation from a Gaussian distribution, and the L_2 computes the distance of decoded volume and input volume in voxel level. VAE is capable of regularizing the shared encoder of the segmentation model.

Our implementation of VAE estimates the mean vector and the standard deviation vector by adding two separate dense layers with 128 output logits. In training, we construct the latent vector by adding mean vector and weighted standard deviation vector by random coefficients in normal distribution. In the validation and testing, we treat the mean vector as the latent vector. To reconstruct the input image, we add a dense layer to recover the shape of input features from the latent vector. With the recovered features, we use trilinear up-sampling and residual convolutional blocks to reconstruct the input image.

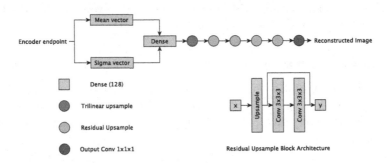

Fig. 3. Model architecture of image reconstruction for variational auto-encoder.

2.5   Training Setup and Implementation

We use batch size 8 with 4 NVIDIA GPUs (Tesla V100 32 GB for client 1 and Quadro RTX8000 for client 2) at each client in all our experiments, and the patches in each batch are randomly sampled and cropped from input volume. The sample rate of foreground patches and background patches are equal. The input patch size we use for training is $[96, 96, 96]$. We use Adam optimizer with learning rate ranged from 10^{-4} to 10^{-5}, with cosine annealing learning rate scheduler. The loss for C2FNAS segmentation is Dice loss combined with categorical cross-entropy loss. In the setting with VAE regularization, we add VAE loss L_{KL} and reconstruction loss L_2 to the total loss with constant coefficients 0.2 and 0.3, respectively.

Our implementation of the C2FNAS model is based on TensorFlow[1]. Our FL experiments utilize the NVIDIA Clara Train SDK[2] for model training and communication of weights between the server and clients.

3   Experimental Results

The experimental setups include standalone training on both clients and federated learning with two clients. In the standalone setting, both Client 1 (**C1**) and Client 2 (**C2**) train their local model independently with each client's

---

[1] https://www.tensorflow.org/.
[2] https://developer.nvidia.com/clara.

own dataset, the resulting models are **C1_baseline** and **C2_baseline**. In the federated learning setup, we set up an aggregation server with no access to any dataset, two clients training on their local datasets sending gradients every ten epochs. The resulting models for federated learning are **FL_global**, **C1_FL_local** and **C2_FL_local**.

Table 1. Dice scores of pancreas and tumor. Data from C1 only have label for panaceas. FL improves the generalizability of model.

Dice coefficient	C1	C2		Pancreas average	Average
	Pancreas	Pancreas	Tumor		
C1_baseline	81.5%	42.4%	0.0%	60.3%	30.1%
C2_baseline	64.7%	65.4%	**54.5%**	65.1%	59.8%
C1_FL_local	81.6%	65.2%	50.2%	72.7%	61.4%
C2_FL_local	81.6%	**66.2%**	52.6%	**73.2%**	**62.9%**
FL_global	**82.3%**	65.4%	46.4%	73.1%	59.8%
Average	78.3%	60.7%	38.3%	68.7%	53.5%

Table 1 compare the standalone training models (C1_baseline and C2_baseline) and FL models (FL_global, C1_FL_local, include C2_FL_local) for C1 dataset and C2 dataset. We have to mention that the C1 dataset only has pancreas label, whereas the C2 dataset is from pancreatic cancer patients, including pancreas and tumor label. For standalone models, the performance is not ideal when predicting on the other client's dataset. C2 tumor even get zero mean Dice socre on C1_baseline model, because the C1 dataset does not include the tumor class. For FL models, the local model, both from C1 and C2, have great improvement on the other dataset. Segmentation performance for tumors on C1_FL_local model is comparable to a standalone model. Even the C1 dataset does not include tumor class. FL_global model shows high generalizability on both C1 and C2 dataset.

Figure 4 shows the qualitative assessment on C1 dataset. When predicting with C2_baseline model, a small region of the pancreas was misdetected as a pancreatic tumor, although CT volumes in the C1 dataset consist of healthy pancreas cases. The misdetection part disappeared after FL. FL_global global model has the best performance on pancreas segmentation for the C1 dataset.

In Fig. 5 we present a visualization of segmentation of one sample volume in the C2 test set. The prediction result of C1_baseline model missed most areas of the pancreas and tumor. The prediction of C2_baseline model is roughly in the correct area, but the shape of the tumor is incorrect and has a false positive of another tumor. The three federated learning models are doing better in detecting the area of the pancreas and the tumor. Although the tumor shape is still far from the ground truth in all predictions, the continuity of the area and the smoothness of the tumor boundary are significantly improved.

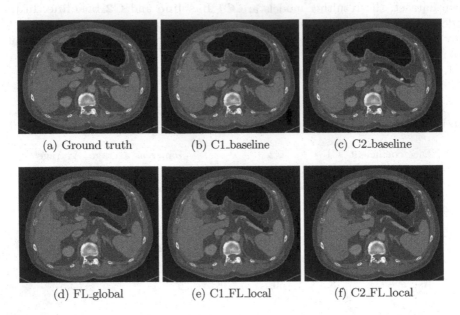

Fig. 4. Comparison of segmentation results with the Client 1 (C1) dataset. Pancreas region in blue and yellow indicates the pancreas and tumor. Only pancreas regions are labeled in the Client 2 (C2) dataset. (Color figure online)

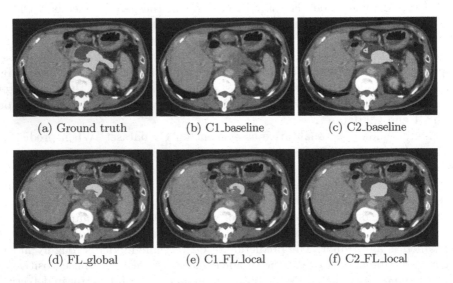

Fig. 5. Comparison of segmentation results with the Client 2 dataset. The pancreas part is labeled as blue and the tumor part is labeled as yellow. (Color figure online)

4 Discussion

In the standalone training setup, both C1_baseline and C2_baseline models perform well on their corresponding local test set. However, the testing results on the opposite test set have a significant performance drop. As the properties of the C1 dataset and C2 dataset are very different (healthy pancreas patients versus patients with pancreatic tumors), it is natural that the standalone models cannot generalize well to different data distribution.

In the federated learning setup, the performance of C1_FL_local model is slightly better than C1_baseline in its own test set, and C1_FL_local model has a remarkable performance gain in the C2 test set, both the mean Dice score of pancreas and tumor on the C2 test set is comparable to the C2_baseline model. For C2_FL_local model, the mean Dice score of the healthy part of the pancreas is slightly better than the C2 baseline model, and the mean Dice score of tumor part drops only moderately. The testing result of C2_FL_local model on the C1 dataset also has a substantial improvement from C2_baseline model, and the performance is similar to C2_FL_local model. The FL_global model can predict well for the pancreas for both test sets, but the prediction of tumors is notably lower than the other two local models. This drop is possibly caused by the lack of any validation or model selection procedure on the server-side. In the client training, we always keep the model with the highest local validation metrics, but on the server-side, the model aggregator only accepts gradients from the clients. The server cannot determine the quality of the model in our current training setting.

5 Conclusions

In this research, we conduct real-world federated learning to train neural networks between two institutes without the need for data sharing between the sites and despite inconsistent data collection criteria. The results suggest that the federated learning framework can deal with highly unbalanced data distributions between clients and can deliver more generalizable models than standalone training.

References

1. Chang, J., et al.: Distributed deep learning networks among institutions for medical imaging. J. Am. Med. Inform. Assoc. **25**(8), 945–954 (2018). https://doi.org/10.1093/jamia/ocy017
2. Li, W., et al.: Privacy-preserving federated brain tumour segmentation. In: Suk, H.-I., Liu, M., Yan, P., Lian, C. (eds.) MLMI 2019. LNCS, vol. 11861, pp. 133–141. Springer, Cham (2019). https://doi.org/10.1007/978-3-030-32692-0_16
3. Li, Z., Hoiem, D.: Learning without forgetting. IEEE Trans. Pattern Anal. Mach. Intell. **40**(12), 2935–2947 (2017)

4. McMahan, H.B., Moore, E., Ramage, D., Hampson, S., y Arcas, B.A.: Communication-efficient learning of deep networks from decentralized data. In: AISTATS (2017)

5. Myronenko, A.: 3D MRI brain tumor segmentation using autoencoder regularization. In: Crimi, A., Bakas, S., Kuijf, H., Keyvan, F., Reyes, M., van Walsum, T. (eds.) BrainLes 2018. LNCS, vol. 11384, pp. 311–320. Springer, Cham (2019). https://doi.org/10.1007/978-3-030-11726-9_28

6. Sheller, M.J., Reina, G.A., Edwards, B., Martin, J., Bakas, S.: Multi-institutional deep learning modeling without sharing patient data: a feasibility study on brain tumor segmentation. In: Crimi, A., et al. (eds.) BrainLes 2018. LNCS, vol. 11383, pp. 92–104. Springer, Cham (2019). https://doi.org/10.1007/978-3-030-11723-8_9

7. Shin, H.C., et al.: Deep convolutional neural networks for computer-aided detection: CNN architectures, dataset characteristics and transfer learning. IEEE Trans. Med. Imaging $35(5)$, 1285–1298 (2016)

8. Tajbakhsh, N., et al.: Convolutional neural networks for medical image analysis: full training or fine tuning? IEEE Trans. Med. Imaging $35(5)$, 1299–1312 (2016)

9. Yang, Q., Liu, Y., Chen, T., Tong, Y.: Federated machine learning: concept and applications. ACM Trans. Intell. Syst. Technol. $10(2)$ (2019). https://doi.org/10.1145/3298981

10. Yu, Q., et al.: C2FNAS: coarse-to-fine neural architecture search for 3D medical image segmentation, December 2019

Fed-BioMed: A General Open-Source Frontend Framework for Federated Learning in Healthcare

Santiago Silva[1(✉)], Andre Altmann[2], Boris Gutman[3], and Marco Lorenzi[1]

[1] Université Côte d'Azur, INRIA Sophia Antipolis, EPIONE Research Project,
Valbonne, France
`santiago-smith.silva-rincon@inria.fr`
[2] COMBINE Lab, Centre for Medical Image Computing (CMIC),
University College London, London, UK
[3] Department of Biomedical Engineering,
Illinois Institute of Technology, Chicago, IL, USA

Abstract. While data in healthcare is produced in quantities never imagined before, the feasibility of clinical studies is often hindered by the problem of data access and transfer, especially regarding privacy concerns. Federated learning allows privacy-preserving data analyses using decentralized optimization approaches keeping data securely decentralized. There are currently initiatives providing federated learning frameworks, which are however tailored to specific hardware and modeling approaches, and do not provide natively a deployable production-ready environment. To tackle this issue, herein we propose an open-source federated learning frontend framework with application in healthcare. Our framework is based on a general architecture accommodating for different models and optimization methods. We present software components for clients and central node, and we illustrate the workflow for deploying learning models. We finally provide a real-world application to the federated analysis of multi-centric brain imaging data.

Keywords: Federated learning · Healthcare · Medical imaging

1 Introduction

The private and sensitive nature of healthcare information often hampers the use of analysis methods relying on the availability of data in a centralized location. Decentralized learning approaches, such as federated learning, represent today a key working paradigm to empower research while keeping data secure [11].

Initially conceived for mobile applications [8], federated learning allows to optimize machine learning models on datasets that are distributed across clients,

Electronic supplementary material The online version of this chapter (https:// doi.org/10.1007/978-3-030-60548-3_20) contains supplementary material, which is available to authorized users.

S. Albarqouni et al. (Eds.): DART 2020/DCL 2020, LNCS 12444, pp. 201–210, 2020.
https://doi.org/10.1007/978-3-030-60548-3_20

such as multiple clinical centers in healthcare [4]. Two main actors play in the federated learning scenario: *clients*, represented for instance by clinical centers, and a *central node* that continuously communicate with the clients [23].

Federated learning methods must address three main issues: *security*, by preventing data leakages and respecting privacy policies such as the EU general data protection regulation (GDPR) [22], *communication efficiency*, by optimizing the rounds of communication between clients and the central node, and *heterogeneity robustness*, by properly combining models while avoiding biases from the clients or adversarial attacks aiming to sabotage the models [1,2]. These issues are currently tackled through the definition of novel federated analysis paradigms, and by providing formal guarantees for the associated theoretical properties [12,13].

Besides the vigorous research activity around the theory of federated learning, applications of the federated paradigm to healthcare are emerging [6,7,10]. Nevertheless, the translation of federated learning to real-life scenarios still requires to face several challenges. Besides the bureaucratic burden, for federation still requires to establish formal collaboration agreements across partners, the implementation of a federated analysis framework requires to face important technical issues, among which the problem of data harmonization, and the setup of software infrastructures. In particular, from the software standpoint, the practical implementation of federated learning requires the availability of frontend frameworks that can adapt to general modeling approaches and application scenarios, providing researchers with a starting point overcoming problems of deployment, scalability and communication over the internet.

In this work we propose Fed-BioMed, an open-source production-ready framework for federated learning in healthcare. Fed-BioMed is Python-based and provides modules to deploy general models and optimization methods within federated analysis infrastructures. Besides enabling standard federated learning aggregation schemes, such as federated averaging (FedAVG) [8,14], Fed-BioMed allows the integration of new models and optimization approaches. It is also designed to enable the integration with currently available federated learning frameworks, while guaranteeing secure protocols for broadcasting.

We expect this framework to foster the application of federated learning to real-life analysis scenarios, easing the process of data access, and opening the door to the deployment of new approaches for modeling and federated optimization in healthcare. The code will be freely accessible from our repository page (https://gitlab.inria.fr/fedbiomed).

2   Related Works

NVIDIA Clara is a large initiative focusing on the deployment of federated learning in healthcare [24], currently providing a service where users can deploy personalized models. The code of the project is not open, and it requires the use of specific hardware components. This may reduce the applicability of federated learning to general use-cases, where client's facilities may face restrictions in the use of proprietary technology.

The Collaborative Informatics and Neuroimaging Suite Toolkit for Anonymous Computation (COINSTAC) [17] focuses on single-shot and iterative optimization schemes for decentralized multivariate models, including regression and latent variable analyses (e.g. principal/independent-component analysis, PCA-ICA, or canonical correlation analysis, CCA). This project essentially relies on distributed gradient-based optimization as aggregating paradigm. A frontend distributed analysis framework for multivariate data modeling was also proposed by [20]. Similarly to COINSTAC, the framework focuses on the federation of multivariate analysis and dimensionality reduction methods. The PySyft initiative [18] provides an open-source wrapper enabling federated learning as well as encryption and differential privacy. At this moment however this framework focuses essentially on optimization schemes based on stochastic gradient descent, and does not provide natively a deployable production-ready framework. Fed-BioMed is complementary to this initiative and allows interoperability, for example by enabling unit testing based on PySift modules.

3   Proposed Framework

3.1   Architecture

Fed-BioMed is inspired by the 2018 "Guide for Architectural Framework and Application of Federated Machine Learning" (Federated Learning infrastructure and Application standard)[1] basing the architecture on a server-client paradigm with three types of instances: central node, clients and federators.

The clients are responsible for storing the datasets through a dedicated client application (Fig. 1, right). Since not every client is required to store the same type of data, this enables studies with heterogeneous or missing features across participants. Moreover, depending on the target data source and on the analysis paradigm, centers can be either included or ignored from the study. Each client verifies the number of current jobs in queue with the central node, as well as data types and models associated to the running jobs.

The central node consists in a secured RESTful API interfacing the federators with the clients, by storing their jobs in queue and submissions. This instance also allows researchers to deploy jobs using predefined models and data.

The federators are in charge of combining the models submitted by the clients for each job, and sharing the global model back through the central node. For this initial stage, Federated Averaging is used as the default federating aggregator. However, as each federator instance is conceptualized as an isolated service, alternative federated aggregators can be included, such as based on distributed optimization via Alternating Direction Method of Multipliers (ADMM) [3]. To allow clients and federators to interact with the RESTful API, we provide a dedicated standard Python library (fed-requests). This eases the procedure for deploying new models or federators into the framework.

---

[1] https://standards.ieee.org/project/3652_1.html.

Communication Scheme: Every instance is packaged and deployed in form of Docker containers [15] interacting between each other through HTTP requests. Containerized instances help to overcome software/hardware heterogeneity issues by creating an isolated virtualized environment with a predefined operating system (OS), thus improving reproducibility as every center run on the same software environment. This scheme also achieves scalability and modularity when dealing with large amounts of clients. In this case, multiple instances of the API can be created under a load balancer, while federator instances can be separately deployed on a dedicated computation infrastructure.

Security: Fed-BioMed addresses typical security issues regarding communication (e.g. man-in-the-middle and impersonation attacks) and access permissions as follows: 1) all requests are encrypted by using HTTP over SSL, 2) user authentication relies on a password-based scheme, and 3) reading/writing operations are restricted by role definitions. Protection from adversarial or poisoning attacks [2] is currently not in the scope of this work, but can be naturally integrated as part of the federator. In the future, malicious attacks will be also prevented by implementing certification protocols attesting the safety of the model source code before deployment [19].

Traceability: to allow transparency to the centers and for the sake of technical support, each instance leverages on a logging system that allows to keep track of every request made, shared data and of the current available jobs.

The architecture behind Fed-BioMed is illustrated in Fig. 1, left. The common procedure involves the deployment of one or multiple jobs from the researchers. Each job must contain the architecture model to be trained and its initialized parameters for reproducibility, the number of rounds, and the federator instance to be used as optimizer.

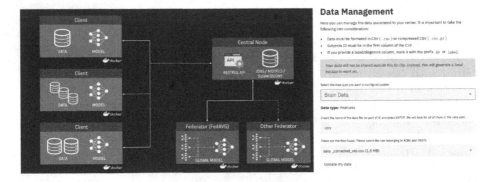

Fig. 1. Left: Clients providing different data sources share their local model parameters with the central node. The central node creates the jobs that will be run by the clients, and transmits the initialization parameters for the models in training. The federator gathers the collected parameters and combines then into a global model that is subsequently shared with the clients for the next training round. As instances are isolated in containers, new instances, such as a new federators (dashed line), can be introduced without altering the behavior of the infrastructure. **Right:** Screenshot of the "Manage Data" section in the client application.

3.2 Workflow

Deploying a New Model. Fed-BioMed relies on a common convention for deploying a new model. A model must be defined as a PyTorch [16] module class, containing the common methods for any torch module:

- __init__(**kwargs) method: defines the model initialization parameters;
- forward() method: provides instructions for computing local model updates.

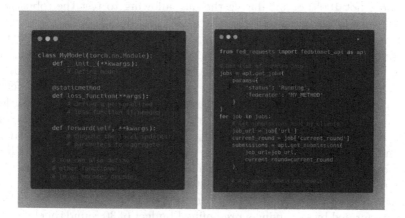

Fig. 2. Examples of model declaration (left) and creation of a new federator (right).

Once the new class is defined, it can be integrated in the model zoo for both clients and federators (Fig. 2, left).

Deploying a new federator or aggregation function in the backend is obtained by creating a containerized service that queries the API for the necessary submissions at each round, and subsequently aggregates the submitted local updates (Fig. 2, right).

4 Experiments

This experimental section illustrates our framework on two different applications: 1) an analysis on the MNIST dataset [9] involving 25 clients, and 2) a multi-centric analysis of brain imaging data involving four research partners based on different geographical locations.

MNIST Analysis: The 60000 MNIST images were equally split among 25 centers. Each center was synthetically emulated and setup in order to interact with the centralized API. The model was represented by a variational autoencoder (VAE) implementing non-linear encoding and decoding functions composed by three layers respectively, with a 2-dimensional associated latent space. For training we used a learning rate of $l = 1 \times 10^{-3}$, 10 local epochs for 30 optimization rounds, while the federated aggregating scheme was FedAVG.

Brain Imaging Data Analysis: In this experiment we use our framework to perform dimensionality reduction in multi-centric structural brain imaging data (MRI) across datasets from different geographical locations, providing cohorts of healthy individuals and patients affected by cognitive impairment.

Four centers participated to the study and were based geographically as follows: two centers in France (Center 1 and Center 4), one in the UK (Center 3), and a last one in the US (Center 2). The central node and the federator were also located in France. Each center was running on different OS and the clients were not 100% online during the day allowing to test the robustness of the framework in resuming the optimization in real-life conditions.

For each center, data use permission was obtained through formal data use agreements. Data characteristics across clients are reported in Table 1. A total of 4670 participants were part of this study, and we note that the data distribution is heterogeneous with respect to age, range and clinical status. 92 features subcortical volumes and cortical thickness were computed using FreeSurfer [5] and linearly corrected by sex, age and intra-cranial volume (ICV) at each center.

This analysis involved data standardization with respect to the global mean and standard deviation computed across centers, and dimensionality reduction was performed via a VAE implementing a linear embedding into a 5-dimensional latent space. Federated data standardization was implemented as in [20], while VAE's parameters aggregation was performed through FedAVG. Federated learning was performed by specifying a pre-defined budget of 30 rounds of client-server iterations in total with 15 epochs/client-round at a learning rate of $l = 1 \times 10^{-3}$.

Table 1. Demographics for each of the centers sharing brain-imaging data. MCI: Mild Cognitive Impairment; AD: Alzheimer's Disease.

	Center 1	Center 2	Center 3	Center 4
No. of participants (M/F)	448/353	454/362	1070/930	573/780
Clinical status				
No. healthy	175	816	2000	695
No. MCI and AD	621	0	0	358
Age ± sd (range) [years]	73.74 ± 7.23	28.72 ± 3.70	63.93 ± 7.49	67.58 ± 10.04
Age range [years]	54–91	22–37	47–81	43–97

5   Results

The evolution of the models during training can be assessed by analyzing the weights' norm across iterations (Fig. 3 and supplementary material for the complete set of weights). MNIST parameters evolution is shown in the top panel, while the related test set projected onto the latent space is shown in Fig. 4, left panel, describing a meaningful variability across digits and samples.

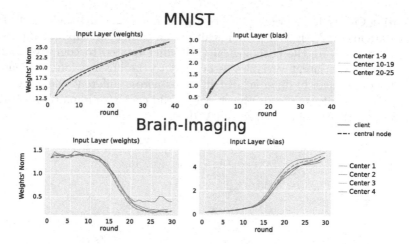

Fig. 3. Illustration of parameter evolution for VAE parameters (input layer). The federated model closely follows the clients' weights distribution. **Top:** MNIST across 25 centers with equally distributed data. **Bottom:** brain-imaging heterogeneous dataset across 4 centers with unevenly distributed data. Continuous lines: clients weights' norm. Dashed lines: federated model weights' norm.

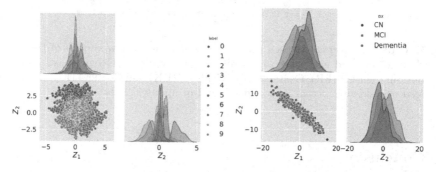

Fig. 4. Left: MNIST pixel data projected onto the latent space. **Right:** Brain features of Center 1 projected onto the first 2 components in the latent space. Although the model was trained with unbalanced data, it is still able to capture pathological variability. CN: healthy controls; MCI: mild cognitive impairment; Dementia: dementia due to with Alzheimer's disease.

Concerning the brain imaging data analysis experiment, the evolution of the encoding parameters throughout the 30 optimization rounds is shown in Fig. 3, bottom panel. For this real-world application we also collected each client's elapsed time to report its local updates to the central node, as well as the average time per round in the best scenario (Fig. 5). As expected, the clients' time varies depending on the geographical proximity with the central node, as well as on the local upload/download speed. The model was further investigated by inspecting the latent space on the subset of the training data available to the coordinating Center 1. The right panel of Fig. 4 shows that although most of the training data for the VAE comes from healthy and young participants, the model is also able to capture the pathological variability related to cognitive impairment in aging. The latent variables associated to the observations of Center 1 indeed show significantly different distributions across different clinical groups, from healthy controls (CN), to subjects with mild cognitive impairment (MCI), and patients with Alzheimer's disease (AD).

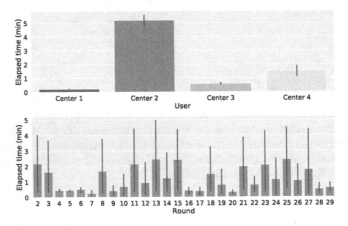

Fig. 5. Top: User average elapsed time per round (since a new version of the model is made available and each user to submit its local update). Geographical and data distribution: Center 1 (FR), Center 2 (US), Center 3 (UK) and Center 4 (FR). **Bottom:** Averaged elapsed time across centers per round of updates.

6   Conclusions

This work presents an open-source framework for deploying general federated learning studies in healthcare, providing a production ready reference to deploy new studies based on federated models and optimization algorithms. Our experimental results show that the framework is stable in communication, while being robust to handle clients going temporarily out of the grid. Scalability is obtained thanks to the use of containerized services. The ability to handle client-authentication and the use of secured broadcasting protocols are also appealing security features of Fed-BioMed. While the experiments mostly focused on VAE and Federated Averaging as aggregating paradigm, our framework is completely extensible to other distributed optimization approaches. Future work will therefore integrate additional models for the analysis of different data modalities and bias, as well as enhanced secure P2P encryption. Concerning the clinical experimental setup, the brain application was chosen to provide a demonstration of our framework to a real-life analysis scenario, and it is not aimed to address a specific clinical question. In the future, the proposed work Fed-BioMed will be a key component for clinical studies tailored to address challenging research questions, such as the analysis of imaging-genetics relationships in current meta-analysis initiatives [21].

Acknowledgements. This work has been supported by the French government, through the 3IA Côte d'Azur Investments in the Future project managed by the National Research Agency (ANR) with the reference number ANR-19-P3IA-0002, and by the ANR JCJC project Fed-BioMed 19-CE45-0006-01. The project was also supported by the Inria Sophia Antipolis - Meediterraneee, "NEF" computation cluster.

References

1. Bagdasaryan, E., Veit, A., Hua, Y., Estrin, D., Shmatikov, V.: How to backdoor federated learning. In: International Conference on Artificial Intelligence and Statistics, pp. 2938–2948 (2020)
2. Bhagoji, A.N., Chakraborty, S., Mittal, P., Calo, S.: Analyzing federated learning through an adversarial lens. In: International Conference on Machine Learning, pp. 634–643 (2019)
3. Boyd, S., Parikh, N., Chu, E.: Distributed Optimization and Statistical Learning via the Alternating Direction Method of Multipliers. Now Publishers Inc. (2011)
4. Brisimi, T.S., Chen, R., Mela, T., Olshevsky, A., Paschalidis, I.C., Shi, W.: Federated learning of predictive models from federated electronic health records. Int. J. Med. Inform. **112**, 59–67 (2018)
5. Fischl, B.: Freesurfer. Neuroimage **62**(2), 774–781 (2012)
6. Gupta, O., Raskar, R.: Distributed learning of deep neural network over multiple agents. J. Netw. Comput. Appl. **116**, 1–8 (2018)
7. Kim, Y., Sun, J., Yu, H., Jiang, X.: Federated tensor factorization for computational phenotyping. In: Proceedings of the 23rd ACM SIGKDD International Conference on Knowledge Discovery and Data Mining, pp. 887–895 (2017)
8. Konečný, J., McMahan, H.B., Ramage, D., Richtárik, P.: Federated optimization: distributed machine learning for on-device intelligence. arXiv preprint arXiv:1610.02527 (2016)
9. LeCun, Y., Cortes, C.: MNIST handwritten digit database (2010). http://yann.lecun.com/exdb/mnist/
10. Lee, J., Sun, J., Wang, F., Wang, S., Jun, C.H., Jiang, X.: Privacy-preserving patient similarity learning in a federated environment: development and analysis. JMIR Med. Inform. **6**(2), e20 (2018)
11. Li, T., Sahu, A.K., Talwalkar, A., Smith, V.: Federated learning: challenges, methods, and future directions. IEEE Signal Process. Mag. **37**(3), 50–60 (2020)
12. Li, T., Sahu, A.K., Zaheer, M., Sanjabi, M., Talwalkar, A., Smith, V.: Federated optimization in heterogeneous networks. arXiv preprint arXiv:1812.06127 (2018)
13. Li, T., Sahu, A.K., Zaheer, M., Sanjabi, M., Talwalkar, A., Smithy, V.: Feddane: a federated newton-type method. In: 2019 53rd Asilomar Conference on Signals, Systems, and Computers, pp. 1227–1231. IEEE (2019)
14. McMahan, B., Moore, E., Ramage, D., Hampson, S., y Arcas, B.A.: Communication-efficient learning of deep networks from decentralized data. In: Artificial Intelligence and Statistics, pp. 1273–1282 (2017)
15. Merkel, D.: Docker: lightweight Linux containers for consistent development and deployment. Linux J. **2014**(239), 2 (2014)

16. Paszke, A., et al.: Pytorch: an imperative style, high-performance deep learning library. In: Advances in Neural Information processing systems, pp. 8026–8037 (2019)
17. Plis, S.M., et al.: Coinstac: a privacy enabled model and prototype for leveraging and processing decentralized brain imaging data. Front. Neurosci. **10**, 365 (2016)
18. Ryffel, T., et al.: A generic framework for privacy preserving deep learning. arXiv preprint arXiv:1811.04017 (2018)
19. Shen, S., Tople, S., Saxena, P.: Auror: defending against poisoning attacks in collaborative deep learning systems. In: Proceedings of the 32nd Annual Conference on Computer Security Applications, pp. 508–519 (2016)
20. Silva, S., Gutman, B.A., Romero, E., Thompson, P.M., Altmann, A., Lorenzi, M.: Federated learning in distributed medical databases: meta-analysis of large-scale subcortical brain data. In: 2019 IEEE 16th International Symposium on Biomedical Imaging (ISBI 2019), pp. 270–274. IEEE (2019)
21. Thompson, P.M., et al.: The enigma consortium: large-scale collaborative analyses of neuroimaging and genetic data. Brain imaging Behav. **8**(2), 153–182 (2014)
22. Voigt, P., Von dem Bussche, A.: The EU General Data Protection Regulation (GDPR). A Practical Guide, 1st edn. Springer, Cham (2017). https://doi.org/10.1007/978-3-319-57959-7
23. Yang, Q., Liu, Y., Chen, T., Tong, Y.: Federated machine learning: concept and applications. ACM Trans. Intell. Syst. Technol. (TIST) **10**(2), 1–19 (2019)
24. Yuhong, W., Wenqi, L., Holger, R., Prerna, D.: Federated Learning powered by NVIDIA Clara (2019). https://devblogs.nvidia.com/federated-learning-clara/

Author Index

Printed in the United States
By Bookmasters